ELECTRICITY AND MAGNETISM

**Recent Titles in
Greenwood Guides to Great Ideas in Science
Brian Baigrie, Series Editor**

Electricity and Magnetism: A Historical Perspective
Brian Baigrie

Evolution: A Historical Perspective
Bryson Brown

The Chemical Element: A Historical Perspective
Andrew Ede

The Gene: A Historical Perspective
Ted Everson

The Cosmos: A Historical Perspective
Craig G. Fraser

Planetary Motions: A Historical Perspective
Norriss S. Hetherington

Heat and Thermodynamics: A Historical Perspective
Christopher J. T. Lewis

Quantum Mechanics: A Historical Perspective
Kent A. Peacock

Forces in Physics: A Historical Perspective
Steven Shore

ELECTRICITY AND MAGNETISM

A Historical Perspective

Brian Baigrie

Greenwood Guides to Great Ideas in Science
Brian Baigrie, Series Editor

Greenwood Press
Westport, Connecticut • London

QC
507
.B35
2007

Library of Congress Cataloging-in-Publication Data

Baigrie, Brian S. (Brian Scott)
  Electricity and magnetism : a historical perspective / Brian Baigrie.
    p. cm. — (Greenwood guides to great ideas in science, ISSN 1559–5374)
  Includes bibliographical references and index.
  ISBN 0–313–33358–0 (alk. paper)
  1. Electricity—History. 2. Magnesium—History. I. Title.
  QC507.B35 2007
  537.09—dc22   2006029542

British Library Cataloguing in Publication Data is available.

Copyright © 2007 by Greenwood Publishing Group, Inc.

All rights reserved. No portion of this book may be
reproduced, by any process or technique, without the
express written consent of the publisher.

Library of Congress Catalog Card Number: 2006029542
ISBN: 0–313–33358–0
ISSN: 1559–5374

First published in 2007

Greenwood Press, 88 Post Road West, Westport, CT 06881
An imprint of Greenwood Publishing Group, Inc.
www.greenwood.com

Printed in the United States of America

The paper used in this book complies with the
Permanent Paper Standard issued by the National
Information Standards Organization (Z39.48–1984).

10 9 8 7 6 5 4 3 2 1

CONTENTS

	List of Illustrations	vii
	Series Foreword	ix
	Introduction	xi
1	Early Investigations	1
2	The Emergence of Systematic Theory	7
3	Electrical Conduction	19
4	Electrostatic Phenomena	29
5	From Effluvia to Fluids	35
6	The Science of Galvanism	49
7	The Current and the Needle	63
8	Forces and Fields	75
9	The Science of Electromagnetism	93
10	Electromagnetic Waves	105
11	Charged Particles of Matter	117
12	The Atom and the New Physics of the Twentieth Century	129
	Timeline	137
	Glossary	145
	Bibliography	153
	Index	159

LIST OF ILLUSTRATIONS

1.1	Diagram of a floating compass	5
2.1	Gilbert's drawing of an instrument of magnetic declination	9
2.2	Gilbert's detector of electrical charge, or versorium	10
2.3	Gilbert's illustration of magnetic variation	13
2.4	Descartes's corpuscular model of magnetic attraction	16
3.1	Boyle's first air pump	20
3.2	Guericke's successful demonstration of the pressure of air	21
3.3	Hauksbee's air pump	24
3.4	Hauksbee's device for rotating amber in a vessel evacuated of air	25
4.1	The Leyden experiment	30
4.2	Experiments with a Leyden jar on animals and plants	32
4.3	Diagram of Volta's electrophorus	34
5.1	Electrical apparatus	38
5.2	Coulomb's torsion balance	43
6.1	Galvani's experiment on frogs	50
6.2	Schematic diagram of Volta's pile	53
6.3	Two forms of Volta's pile	54
7.1	Drawing of Oersted's experiment	66
7.2	Diagram of Ampère's experimental arrangement	70
7.3	Ohm's experimental arrangement	73
8.1	Experimental apparatus for Faraday's rotations	76
8.2	Faraday in his physics laboratory at the Royal Institution	79
8.3	Illustration of Arago's disk	81
8.4	Faraday's apparatus for electromagnetic induction	82
8.5	Illustration of Faraday's disk	83
8.6	Faraday's apparatus for electrolytic decomposition	85

List of Illustrations

9.1	Array of vortices (end on)	98
9.2	Drawing of the disturbance of a magnetic field by an electric current in a straight conductor	100
10.1	Diagram of Hertz's apparatus	106
10.2	Hertz's map of electromagnetic waves	107
11.1	Crookes tube	119
11.2	Diagram of J. J. Thompson's apparatus for showing that cathode rays are negatively charged	125
12.1	Diagram of apparatus for R. A. Millikan's oil-drop experiment	134

SERIES FOREWORD

The volumes in this series are devoted to concepts that are fundamental to different branches of the natural sciences—the gene, the quantum, geological cycles, planetary motion, evolution, the cosmos, and forces in nature, to name just a few. Although these volumes focus on the historical development of scientific ideas, the underlying hope of this series is that the reader will gain a deeper understanding of the process and spirit of scientific practice. In particular, in an age in which students and the public have been caught up in debates about controversial scientific ideas, it is hoped that readers of these volumes will better appreciate the provisional character of scientific truths by discovering the manner in which these truths were established.

The history of science as a distinctive field of inquiry can be traced to the early seventeenth century when scientists began to compose histories of their own fields. As early as 1601, the astronomer and mathematician, Johannes Kepler, composed a rich account of the use of hypotheses in astronomy. During the ensuing three centuries, these histories were increasingly integrated into elementary textbooks, the chief purpose of which was to pinpoint the dates of discoveries as a way of stamping out all too frequent propriety disputes, and to highlight the errors of predecessors and contemporaries. Indeed, historical introductions in scientific textbooks continued to be common well into the twentieth century. Scientists also increasingly wrote histories of their disciplines—separate from those that appeared in textbooks—to explain to a broad popular audience the basic concepts of their science.

The history of science remained under the auspices of scientists until the establishment of the field as a distinct professional activity in the middle of the twentieth century. As academic historians assumed control of history of science writing, they expended enormous energies in the attempt to forge a distinct and autonomous discipline. The result of this struggle to position the history of science as an intellectual endeavor that was valuable in its own right, and not merely in consequence of its ties to science, was that historical studies of the natural sciences were no longer composed with an eye toward educating a wide audience that included non-scientists, but instead were composed with the aim of being consumed by other professional historians of science. And as historical breadth was sacrificed for technical detail, the literature became increasingly daunting in its technical detail. While this scholarly work increased our understanding of the nature of science, the technical demands imposed on the reader had the unfortunate consequence of leaving behind the general reader.

As Series Editor, my ambition for these volumes is that they will combine the best of these two types of writing about the history of science. In step with the general introductions that we associate with historical writing by scientists, the purpose of these volumes is educational—they have been authored with the aim of making these concepts accessible to students—high school, college, and university—and to the general public. However, the scholars who have written these volumes are not only able to impart genuine enthusiasm for the science discussed in the volumes of this series, they can use the research and analytic skills that are the staples of any professional historian and philosopher of science to trace the development of these fundamental concepts. My hope is that a reader of these volumes will share some of the excitement of these scholars—for both science, and its history.

Brian Baigrie
University of Toronto
Series Editor

INTRODUCTION

This volume attempts to describe the main developments of the science of electricity and magnetism from the early modern period, when lore that had flourished unchecked for more than two millennia was displaced by a body of systematic theory, to the first decade of the twentieth century when Maxwell's electromagnetic theory was replaced by Einstein's special theory of relativity. It is my hope that this volume will be of interest to the specialist and the general reader alike.

Admittedly, an exhaustive history of this subject is a lofty and, regrettably, unrealizable goal. The sheer volume and variety of scientific publications on electricity and magnetism is daunting, especially when one traces the development of the science of electricity and magnetism through to its application to problems in engineering. To add to this mountain of historical documentation, it must be acknowledged that the science of electricity and magnetism, which exploded in the first decades of the nineteenth century, did not give us just a new and vibrant field of physics, with fruitful ties to such scientific fields as chemistry, astronomy, and biology, but an entirely new kind of civilization, in less than a hundred years. Clearly, a comprehensive study would have to address the enormity of these far-reaching social consequences if it is to do justice to the science.

To make the subject matter more manageable, my narrative focuses on concept formation and novel device production. Technological applications, even those that were revolutionary in their social ramifications, are only mentioned in passing, if at all, unless an application engendered a feedback on the conceptual and instrumental development of the field. Despite this trimming, I have incorporated in this volume two postscripts that give the reader an overview of applications associated with power and communication technologies. Even in these postscripts, though, the emphasis has been on the creation of the science itself and not its broader social significance.

The reader will note that I have selected for most detailed discussion those transitional periods that witnessed the introduction of novelty—novel theories, experiments, and device production. My rationale for this emphasis is that the creation of a body of theory and associated experimental practice, rather than the articulation of theoretical frameworks, lends itself to nontechnical presentation. A detailed recounting of the score of experimental tests at play in certifying this or that theory as knowledge is inevitably daunting in its technical nature, and it becomes more so as our story approaches the twentieth century and physics is increasingly swamped by technology.

The consequence of this decision to focus on the production of novelty is that a great deal of emphasis has been placed on the individual scientists who transformed the foundations of the science of electricity and magnetism through their theoretical and experimental efforts. The result is that many institutions and organizations—for example, the Royal Society, the Académie Royale des Sciences, and the Royal Institution—may seem to receive comparatively less attention, perhaps striking some readers as running contrary to the usual contemporary humanistic ruminations on the development of modern science that give pride of place to social organization.

The relationship between mathematics and theoretical innovation is often critical to the fortunes of a potentially fruitful line of inquiry. The reader may therefore be taken aback to discover that I have elected, as much as possible, not to explore the details about mathematical technicalities. Although conceding that this decision risks leaving the reader with a distorted overall picture of physics, I believe the broader perspective adopted in this volume has many advantages, the chief being that it helps to highlight large-scale changes in method and shifts in disciplinary boundaries, which tend to vanish when the period of the historical account spans the life of a single central figure or a particular experimental device. Numerous well-crafted historical studies are devoted to the technical details of the science of electricity and magnetism, but less than a handful trace the history of scientific ideas and their broader cultural and philosophical meanings.

This philosophical gaze has also allowed me to describe the emergence of the central achievements in a manner that is sensitive to the interconnections between different scientific niches. The path of the science of electricity and magnetism from the early modern period to the twentieth century often winds its way through other scientific domains, through anatomy (in the case of Galvani), chemistry (in the case of Davy and Faraday), and the theory of heat (in the case of Maxwell).

Although I have kept formalisms to a minimum, sometimes the available mathematics constrained conceptual developments and other times new representations placed a premium on mathematical innovation. In these cases,

I have attempted to render these technical matters accessible to the intended audience of this volume. And, wherever possible, I have sought to respect the original styles of mathematical demonstration, unless otherwise noted.

This volume has evolved from lectures on the history of the natural sciences that I have delivered over the past 15 years. I have benefited from numerous conversations with my colleagues (past and present) at the Institute for History and Philosophy of Science and Technology—especially Trevor Levere, Sungook Hong, Jed Buchwald, and Thaddeus Trenn. I want to thank these colleagues, as well as the many undergraduate students over the years for scores of questions and requests for clarification that fueled my continuing interest in the subject matter of this book.

I want to thank the Thomas Fisher Rare Book Library at the University of Toronto for permission to reproduce many of the illustrations found in the text. My appreciation also goes to the staff at Fisher for their help and guidance. Special thanks to Jeff Dixon for original artwork.

1

EARLY INVESTIGATIONS

LORE ABOUT AN ANCIENT MYSTERY

Thales (c. 640–546 B.C.) is widely acknowledged by historians as the first philosopher of the Greek tradition. On the authority of a statement in Aristotle's *De Anima* (On the Soul) (405a19), he is also frequently credited with first reporting an observation upon which the edifice of the science of electricity has been erected, namely, that after amber, a fossil resin much esteemed for the manufacture of ornamental objects, has been rubbed with cloth, it attracts small pieces of wool, hair, lint, and other light objects. Indeed, our modern word *electricity* is derived from the Greek word for amber — *electron*. The Greeks and Romans were not the only ancient people who were acquainted with the attraction of rubbed bodies for chaff and paper; Chinese and Persian investigators recorded the phenomenon of electrical excitation as well. This class of phenomena is now known as static electricity and its study electrostatics.

Static electricity occurs, as we now know, when two dissimilar insulating materials are placed into intimate contact and then separated. All that is required is the touching; rubbing is not necessary. If one of the two materials is rough or fibrous, it does not produce a very large contact area, and so rubbing one material upon another can augment the contact area; but this rubbing is not the cause of the electrification. Chemical bonds are formed when the surfaces touch. If the atoms in one surface tend to embrace electrons more tightly, this surface will tend to appropriate charged particles from the other surface the instant they touch. This appropriation, in turn, causes the surfaces to become oppositely "charged," so that they acquire imbalances of opposite polarity. One surface will now possess more electrons than protons, while the other will possess more protons than electrons. If the surfaces are subsequently separated, the regions of opposite-charge imbalance will also separate.

Another class of attractive phenomena that did not escape the attention of the ancients was that involving iron and iron ore. Fragments of a certain naturally occurring type of iron were found to attract iron and nothing else, as nearly as the ancients could tell, for no reason that was immediately apparent. Such a fragment was called a lodestone (sometimes spelled loadstone). The iron ore that we now call magnetite is common in nature, but the fragments of iron ore possessing the attractive virtue were exceedingly rare. Just why a few pieces of magnetite differed from the rest was a mystery. Another mystery was that the attractive power of the lodestone was strikingly similar to that of amber, but lodestone worked its magic without its first having to be rubbed.

The Greeks and Romans were also aware that the lodestone not only attracted iron but also invested iron with its own magnetic property; that is, when in contact with a lodestone, an iron ring would hang from it and, in turn, a second iron ring would hang from the first, and so on, until a certain number was reached, at which point the attractive power would be overcome by the weight of the iron rings. The ancients believed that the attraction was confined to iron and that it was not destroyed when other bodies were placed between the magnet and iron, but they were mistaken on this score. Iron, some steels, and the mineral lodestone exhibit easily detectable magnetic properties, but all materials are influenced to some degree by the presence of a magnetic field. In most cases, however, it is difficult to detect this influence without delicate instruments.

Thales lived in the town of Miletus in northern Greece on the outskirts of Greek civilization. The sample of iron ore that he examined was said to have come from the neighboring town of Magnesia. Thales called his sample "the magnesian rock." Such iron-attracting substances are now called *magnets*, and the phenomenon itself is called *magnetism*.

According to Aristotle, Thales also developed an account of the attractive virtue of the magnet: "Thales, too, to judge from what is recorded about him, seems to have held soul to be a motive force, since he said that the magnet has a soul in it because it moves the iron" (*De Anima* 405a19). Thales's reasoning, if Aristotle's report is accurate, was based on an analogy between the soul that was believed to animate and thereby move living things, and the attractive virtue of the magnet: Because souls cause the motions of living things, inasmuch as the lodestone moves iron, it must be because it is animated by a soul of some sort. The behavior of the lodestone, in other words, is to be explained in terms of an *immaterial* force of attraction that is concealed from sensation but that nevertheless acts between bodies situated a distance apart from one another.

In contrast to Thales, Empedocles of Akragas (491–435 B.C.) advanced an account of the lodestone's attraction that is *mechanical* in character. It is mechanical in that the iron is attracted by the stone, in Empedocles's view, simply because it comes into contact with it. Indeed, he believed that without this contact between the iron and the stone, there would be no magnetic influence. According to Alexander of Aphrodisias (fl. A.D. 200), a philosopher well known for his commentaries on the chief works of Aristotle, Empedocles attempted to explain the lodestone's attraction by the postulation of *effluvia* or tendrils emanating from both the iron and the stone. According to his account of magnetic attraction, the effluvium of the stone first expels the air from the pores of the iron. Once expelled, the iron itself is carried along by the abundant flow of the effluvium. Again, when the effluvium of the iron moves toward the pores of the stone, which are fitted with pores to receive it, the iron follows and moves with it. This theory had much to recommend it. In particular, it explained why iron alone was affected by the emanations; that is, the pores were fitted only to the iron, and not other materials, and so acted only on iron.

These two theoretical approaches—Thales's appeal to an immaterial cause that acts at a distance and Empedocles's materialistic theory in the form of insensible effluvia that act by direct contact—are the two positions that dominated thinking about electricity and magnetism well into the nineteenth century. As we shall see in the coming chapters, the many theories that were devised over these centuries can be seen as modifications of one or other of these two intellectual approaches. Although both positions, strictly speaking, appeal to underlying causes that are concealed to sensation, only the latter theory of Empedocles appeals to a cause that is nonetheless material. The immaterial theory of Thales, in contrast, in the coming centuries would be branded as an occult quality.

During the course of many centuries, the powers invested in the lodestone magnified many times over. The lodestone was credited with all sorts of powers—to cure dropsy, mange, and burns, and conversely, to cause melancholy (Ennemoser 1854, 2:334–335). It was credited with attracting and dissolving ships made of iron. The lodestone was even touted as a means for exposing unfaithful spouses.

THE FIRST EXPERIMENTAL SCIENTIST

Around the year A.D. 1000, the Chinese discovered that when a lodestone or an iron magnet was placed on a float in a bowl of water, it always pointed north and south. It was this discovery that led to the invention of the magnetic compass, which quickly spread to the Arabs and from the Arabs to Europe. The compass, which was in general use in Western ships by the twelfth century, helped ships navigate safely in the open sea, even when clouds covered the stars.

A noteworthy exception to this unbridled speculation about the powers of the lodestone was the work of Petrus Peregrinus of Maricourt (fl. thirteenth century). Little is known about the life and work of Peregrinus. What is known is that in 1269, during a tour of service in the army of Charles of Anjou, King of Sicily, he composed a letter that described everything known about the lodestone. His letter, or *Epstola de Magnete* (*Epistle on the Magnet*) as it is now known, even included detailed instructions (complete with diagrams) for the construction of two kinds of magnetic compasses.

The letter is divided into two sections. The first section is a discussion of the facts of magnetism in light of his experiments. The second section is a description of two instruments for the directional measurement of celestial bodies and a machine of perpetual motion based on magnetic energy.

Taking a natural magnet that had been rounded into a globular form, Peregrinus placed it on a needle and noted the line along which the needle set itself. He then laid the needle on other parts of the stone, obtaining more lines in this manner. He found that the lines formed circles, which girdled the stone in much the same manner as lines of longitude girdle the Earth. Just as the lines of longitude pass through the arctic and antarctic poles of the Earth, the lines passed through two points at opposite ends of the lodestone. He called these centers of attraction "poles" of the magnet.

Until Peregrinus, no one had obtained any clear understanding of the polarity of the magnet. It was known in a vague way that some parts of the lodestone would attract iron more strongly than other parts and that iron was apparently repelled under other conditions. Peregrinus noted that the way in which magnets set themselves and attracted each other depended on the position of their poles, as though these were the seats of magnetic powers. He also found that, under the influence of a stronger stone, it was possible to change the magnetic polarity of a lodestone. When a magnet was cut in two, each portion was thereby made into a magnet, complete with its own north and south poles. Peregrinus dedicates the better part of a chapter to this discovery, detailing the correct manner of placing the pieces together in the right order so that they form a single magnet once more.

To explain the propensity of lodestones to point north and south, Peregrinus advanced the intriguing cosmological hypothesis that an invisible celestial pole imparted its power to the poles of a magnetized needle, as well as to every part of a spherical lodestone. In chapter 5, Peregrinus advances the following argument in support of this hypothesis:

> Since the loadstone points to the south as well as to the north, it is evident from the foregoing chapters that we must conclude that not only the north but also from the south pole rather than from the veins of mines virtue flows into the poles of the loadstone. This follows from the consideration that wherever a man may be, he finds the stone pointing to the heavens in accordance with the position of the meridian; but all meridians meet in the poles of the world; hence it is manifest that from the poles of the world the poles of the loadstone receive their virtue. Another necessary consequence of this is that the needle does not point to the pole star but to the poles of the world ... From all these considerations, it is clear that the poles of the loadstone derive their virtue from the poles of the heavens (1269/1902).

Peregrinus's claim seems to be that the virtue of the lodestone is obtained from the heavens; else they would not point toward them. It is hardly surprising that, at a time when the stars and the planets were widely believed to hold sway over human affairs, Peregrinus would be held captive by this belief, but is it true that the loadstone points to the celestial poles at all times and all places? The discovery that the compass needle sometimes points away from the poles would have been conclusive proof that his theory was misguided, but Peregrinus never performed the experiments that could have brought this fact to light.

To sustain his conviction that the poles of the magnet receive the virtue from the poles of the universe, Peregrinus conceived an ingenious thought experiment that was firmly erected on his acceptance of the reigning Ptolemaic cosmology, in which the heavens were held to move around a stationary Earth. Because the behavior of the stone is governed by the poles of the sky, and because the sky turns round its poles once every 24 hours, a well-balanced globular magnet mounted on frictionless bearings, Peregrinus insists, will also turn round its axis in unison with the heavenly sphere. Peregrinus attributed his inability to detect such a rotation to the lack of a suitable device to detect this motion, and not to a flaw in nature or in his chain of reasoning.

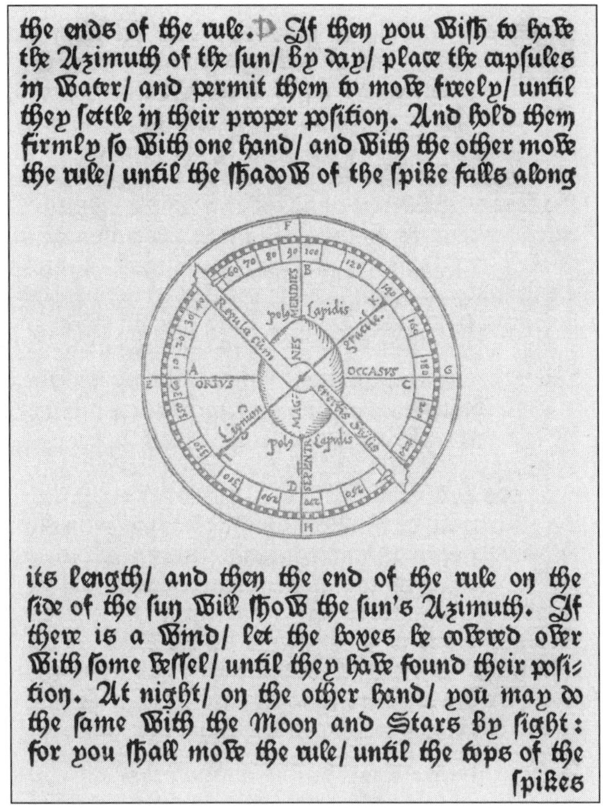
the ends of the rule. If then you wish to have the Azimuth of the sun/ by day/ place the capsules in Water/ and permit them to move freely/ until they settle in their proper position. And hold them firmly so with one hand/ and with the other move the rule/ until the shadow of the spike falls along its length/ and then the end of the rule on the side of the sun will show the sun's Azimuth. If there is a wind/ let the boxes be covered over with some vessel/ until they have found their position. At night/ on the other hand/ you may do the same with the Moon and Stars by sight: for you shall move the rule/ until the tops of the spikes

Figure 1.1: Diagram of a floating compass. From Peregrinus, *Epistle of Petrus Peregrinus of Maricourt* (1269/1902).

In the second part of his letter, Peregrinus details the construction of three types of magnetic instruments. The first is a floating compass, discussed earlier in this chapter. Peregrinus's compass design represented a significant improvement on extant compasses by the addition of a fiducial line and a circle divided out with gradations—modifications that provided the sailor with marks by which the helm could be set (fig. 1.1). The second instrument is a pivoted compass, featuring a magnetized needle that is thrust through an axis set between two sets of pivots, and a transverse index set at right angles to the axis. When the needle itself lies north and south, the transverse axis will lie east and west. Peregrinus also describes a transparent cover and a sighting method by means of which "you may direct your course toward cities and islands and all other parts of the world, either on land or at sea, provided you know beforehand the longitudes and latitudes of those places (1269/1902)." The third instrument is a new wheel of perpetual motion, which is intriguing but does not concern us here.

Peregrinus's efforts were so remarkable that in his day he was conferred the title of the world's first experimental scientist by Roger Bacon (1220–1292), himself a pioneer of the experimental method who taught at the universities of Oxford and Paris. "He is a master of experiments," Bacon wrote in chapter 13 of his *Opus Tertium* (1287: Third Work), "and thus by experience he knows natural, medical, and alchemical things, as well as all things in the heavens and beneath them." (qtd. in Grant 1981, 533). What is most intriguing about Bacon's portrait of Peregrinus as a pioneer of the experimental method is that it was drawn two years before Peregrinus wrote his immortal letter on magnetism, suggesting that he had written with great authority on other scientific matters. Unfortunately, only his letter on the magnet survived until modern times.

In its day, Peregrinus's letter was not well known. It was copied a number of times and fairly widely distributed in libraries throughout Europe: there are 31 extant copies in manuscript form—29 in Latin and 2 in Italian. Five others that were thought to exist have disappeared. Although the original letter was copied many times, it does not follow that it was well known. Indeed, for 300 years the letter was completely forgotten until it was published in 1558. The fact that there are no references to Peregrinus's letter in any works on magnetism until the publication of Gilbert's *De Magnete* (1600) supports the view that the letter was known only to librarians.

Although a number of tantalizing references to electrical and magnetic phenomena appear in the decades following the publication of Gilbert's work, 300 years would pass before these phenomena were once again given careful study. Indeed, the two notable discoveries of the Renaissance related to the science of electricity and magnetism—namely, the inclination and the declination of the magnetic needle—were made by sailors and not by natural philosophers.

2

THE EMERGENCE OF SYSTEMATIC THEORY

INTRODUCTION

With the remarkable exception of Peregrinus, the first turning point in the study of electrical and magnetic phenomena occurred in the latter half of the sixteenth century, when natural philosophers discovered means of criticizing lore that had flourished unchecked for more than two millennia. With the rise of new experimental methodologies and associated experimental technologies, scientists were at last in a position to confront nature directly and so impose a check on unbridled speculation. Experiment allowed scientists to move from general sorts of questions to specific questions about the details surrounding the production of these striking phenomena.

As we shall see, however, this movement from general sorts of questions to more detailed experimental investigations of electrical and magnetic phenomena did not amount to a scientific revolution in the sense of the expression readily associated with the paradigm names of Newton, Lavoisier, and Darwin. Problems in optics, mechanics, and astronomy—those sciences with a discernible mathematical structure—stood at the forefront of investigation of natural philosophers during the seventeenth century. The study of electricity and magnetism, in contrast, was bereft of a mathematical core—a deficit that would not be eradicated until the mid-eighteenth century with the work of Coulomb and others. Although experiment exerted an important control on the production of theory, thereby distinguishing the unbridled speculation of the ancient philosophers from the natural philosophers of the early modern period, the absence of mathematical analysis curtailed the articulation of the science of electricity and magnetism. The result was that the theory fashioned by the central figures of the seventeenth century, while true to the facts of the matter as established by experiment, did not move in a meaningful way beyond the theoretical approaches associated with Thales and Empedocles.

THE EARTH AS A MAGNET

The systematic study of electricity and magnetism began in the late sixteenth century with the researches of William Gilbert of Colchester (1544–1603), appointed in 1601 as court physician to Elizabeth I. The queen died two years later from the plague, which periodically afflicted London during the seventeenth century. Gilbert succumbed to the plague shortly after the Queen..

Gilbert gathered together all known opinions concerning the magnet in a pioneering volume that was the culmination of seventeen years of intense labor and research. This treatise was titled *De Magnete* (1600: On the Magnet: Magnetic Bodies Also, and On the Great Magnet the Earth; a New Physiology, Demonstrated by Many Arguments and Experiments). As the extended subtitle attests, Gilbert advances the remarkable thesis that the Earth itself is a huge magnet, or *terrella*, and to this extent, at least, subject to the same laws as an ordinary magnet:

> The lodestone derives properties from the earth … magnetic bodies are governed and regulated by the earth, and are subject to the earth in all their movements. All the movements of the lodestone are in accord with the geometry and form of the earth and are strictly controlled thereby … Such, then we consider the earth to be in its interior parts; it possesses a magnetic homogenic nature. (Gilbert 1600, 216).

De Magnete is an important work, not merely as a consequence of this remarkable thesis that the Earth itself is a great magnet, but also in consequence of the manner in which it marshals the results of experiment and observation to sustain this thesis. In this respect, at least, Gilbert's treatise is the first work in science that is unabashedly modern. The manner in which Gilbert invokes experiment both to interrogate opinions about the behavior of the magnet and to sustain his novel magnetic philosophy was also bold in 1600, a time when experimental practice was still widely regarded with suspicion. Indeed, if an experimental result was seen to conflict with received opinion, it was not uncommon for the source of the result to be accused of deception. During the course of the seventeenth century, Galileo, Descartes, Newton, and other scientists developed a number of strategies to minimize the perceived role of experiment in their deliberations.

Still, relatively few of Gilbert's experimental discoveries originated with him. Many were first reported by Peregrinus and Robert Norman (c. late sixteenth century), who was the most important English contributor to magnetic science prior to Gilbert. Norman was a retired sailor, who had taken up the profession of constructing instruments of navigation. He used a compass needle, balanced on a horizontal axis, which was able to swing in a vertical plane lined up in the north-south direction. He noticed that the needle did not come to rest in the horizontal plane but inclined slightly downwards. To counter this "dip angle," Norman snipped off pieces of his magnetized needle that inclined downward because it appeared to weigh more. He found that balance could only be achieved when weight was added to the end opposite the one that inclined downward. He also found that he was only able to balance an unmagnetized needle. On being magnetized, the needle would incline downward once more.

This discovery that the force on a magnetic needle was not horizontal but pointed down toward the Earth (fig. 2.1) was published by Norman in 1581 in *The Newe Attractive*—the first book in English devoted to the study of magnetic phenomena. (This phenomenon of the *magnetic dip* had been discovered previously by George Hartmann in 1544, but his letter describing his discovery was not well disseminated.) The inference that seems obvious is that it was magnetism in the Earth that caused the dip of the needle, but Norman did not come to this conclusion. It was left to Gilbert to draw the inference that the Earth is the source of the magnetic dip.

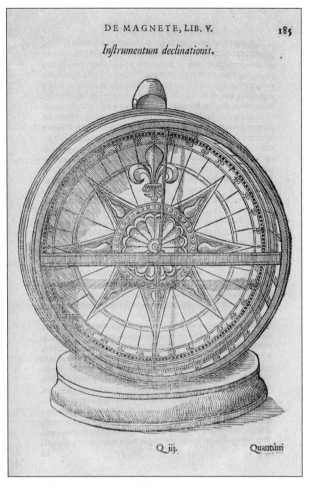

Figure 2.1: Gilbert's drawing of an instrument of magnetic declination. From Gilbert, *De Magnete* (1600). With kind permission of the Thomas Fisher Rare Book Library, the University of Toronto.

Some scholars have questioned whether Gilbert came to his central idea unassisted. In 1597, William Barlowe published a treatise on magnetism called *The Navigator's Supply*. Barlowe was the first to refer in print to "the magnetism of the Earth" as the source of magnetic phenomena. Because Gilbert and Barlowe were intimately acquainted, it is extremely difficult to ascertain which of these two had the idea first. In Gilbert's defense, he did not claim that the central ideas and results of his great work were his alone. His sources are not credited as fully as one would like, but experimental results are marshaled in a textbook manner to sustain the magnetic philosophy. Indeed, advancing the magnetic philosophy, rather than laying claim to particular discoveries, is the overriding purpose of *De Magnete*. In this respect, this treatise should be viewed as a *repository* of magnetic knowledge at the beginning of the seventeenth century.

Gilbert's monumental work is divided into six books, each with further subdivisions into chapters. The first book addresses the lodestone and the history of its study; the second book discusses the five magnetic movements, especially the phenomenon of the magnetic attraction of unlike poles.[1] Chapter two of the second book tackles the issue of the amber effect in consequence of its similarity to magnetism. In the remaining books, theories of magnetic and electric attraction are presented and contrasted: the third book introduces the directive magnetic movement, including the modern concept of polarity; the fourth introduces the theory of compass variations as applied to navigation; the fifth discusses the magnetic dip and its measurement; and the sixth continues Peregrinus's experiments with spherical magnets, concluding that the spherical magnet (*terrella*, or small Earth) is a model of the Earth.

The first book opens with a discussion of the opinions of the ancients and of Gilbert's immediate Renaissance predecessors regarding the attractive powers of the lodestone and electric bodies. Gilbert reveals himself to be well versed in the theories of antiquity and of the middle ages. He is especially taken up with the mechanical theories of the Greek philosophers, Gilbert's objections to these theories accompanied, at times, by astute experiments that are designed to put them to the test. For example, the theory attributed to Hippocrates (c. 460–377 B.C.), which explains attractions in terms of heat, is shown by Gilbert to be groundless by the heating of a lodestone that is seen to lose its attractive virtue. The ancient atomist hypothesis that electric and magnetic attractions take place by the motion of air and resulting suction is refuted, in the case of magnetism, by showing that nothing corporeal passes between the lodestone and iron and, in the case of electrical attraction, by an ingenious experiment with a candle flame that is not disturbed by the presence of a strongly attracting piece of amber, as one would expect if the amber produced a current of air. The Aristotelian attempt to explain magnetism in terms of the interplay of four fundamental elements (earth, air, fire, and water) and the prime qualities (hot, dry, wet, and cold), Gilbert says, is to be relinquished to "the moths and the worms."

The second book contains Gilbert's electrical work, the first ever published on this subject. This book of *De Magnete* marks the beginning of the science of electricity. In it Gilbert describes the well-known attractive property of amber. An "electric," he claims, is a substance that attracts when rubbed—except sometimes it may attract without being rubbed, if it is both finely polished and a large piece of a very good electric. He also remarks that a large number of substances other than amber—for example, diamond, sapphire, rock crystal, glass, sulfur, sealing wax, and resin—also possess this property, thereby exploding the opinion, which had prevailed for more than 2,000 years, that the attractive power of amber is a property of this particular substance. In this way, Gilbert transformed the Greek word for amber into a working

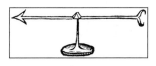

Figure 2.2: Gilbert's detector of electrical charge, or versorium. From Gilbert, *De Magnete* (1600). Courtesy of the Thomas Fisher Rare Book Library, the University of Toronto.

definition that would form the basis of future electrical research. Gilbert also discusses a class of substances—pearls, marble, and agate—that exhibit no sign of electrification when held in the hand and rubbed. He calls these materials "non-electrics"—an erroneous expression because if these substances are fastened to a glass handle and then rubbed, they behave as electrics. He then produces a list of electrics and a list of non-electrics.

In the second book, we are also introduced to Gilbert's *versorium* (fig. 2.2), a detector of electrical charge that Girolamo Fracastoro (1478–1533) had utilized to show that amber draws silver as well as chaff. This instrument consisted of a stiff straw balanced lightly upon a sharp point. When an electrified body was held near the pivoted electroscope, it was attracted and turned around. The versorium, crude as it was, could detect the presence of amounts of electricity far too small to attract bits of paper from a table. More than a century would pass before a concerted effort was made to improve this device.

Although he rejects on experimental grounds the atomist hypothesis that electric attraction is due to the motion of air, Gilbert dusts off Empedocles's theory of effluvia, suggesting that a rubbed electric exudes an effluvium, or stream of small particles, that flows out of bodies without any sensible diminution of the bulk or weight of the body that emits them. This effluvium acts directly on the attracted body and can be stopped by a thin membrane such as linen, by a flame, or by moisture. Gilbert did not record electrical repulsion, nor could he formulate electrical conduction, because he believed that an object in contact with the excited electric suppressed the effluvium. This theory of a material electrical effluvium continued to exert great influence throughout the seventeenth and eighteenth centuries and, as we shall see in later chapters of this book, it was the precursor of the notion of an "electric fluid" developed by Dufay, Nollet, and Franklin and the concept of "electric charge."

Turning to magnetic attraction, if there are magnetic effluvia similar to electric effluvia, Gilbert contends that presumably they must be capable of penetrating iron because the magnetic power can be transferred from lodestone to iron. But Gilbert points to experiments that show that iron is affected even through thick screens of dense matter, through which electric attraction cannot penetrate. On this basis, he concludes that magnetism cannot be conveyed by material particles, as is the case with electric attraction. Magnetic attraction, instead, is due to a "form" that is more concentrated in lodestone because this material is more earthlike than other materials. As an explanation of the fact that only some fragments of magnetite possess the magnetic virtue, Gilbert argued that lodestones were created by long-term exposure to the Earth's magnetism; that is, the poles of such lodestones were turned to the poles of the Earth for a long period of time.

The ancients had associated the attractions of amber and the lodestone, accounting for both in the same general manner. In the mid-sixteenth century, the mathematician Girolamo Cardano (1501–1576) had drawn up a catalog of differences between magnetic and electric phenomena. Gilbert was aware of this list, and doubtless was inspired by it. In step with Cardano, he drew a distinction between electrical and magnetic excitation, but then took the additional step of stipulating that neither constitute "attractions" in the appropriate sense of the term. If our frame of reference for the attraction is the "dragging along" that Gilbert and his contemporaries associated with a horse pulling a cart, it is reasonable to assume with Gilbert that attractions called for coupled motion and violence. Electrical action is violent but not attractive; that is, unlike the cart, the electric does not move. In keeping with his belief that the electric does not move, Gilbert preferred the term "incitation" to describe the amber effect. Magnetic action, on the other hand, is mutual but nonviolent — that is, it is a coition or coming together.

Of course, it is difficult for us now to conceive that a horse pulling a cart is doing violence to the cart unless, that is, we presume with Gilbert that the cart and other ponderable bodies at rest are thereby fulfilling their natures, and that putting them in motion is a violation of their natures. This idea that motion and rest are opposed to one another was a guiding assumption of the Aristotelian physics that dominated intellectual life in the West until it was abandoned in the latter half of the seventeenth century in favor of the law of inertia, which stipulated that motion and rest alike are *states* of matter.

Gilbert is not suggesting, then, that the Earth is magnetic because it contains magnetic substances (a view that is frequently attributed to him). Rather, his view is that those substances are magnetic that, by their very nature, are intrinsically terrestrial. The terrestrial poles are not mere geometrical points, as was widely believed, but physical points. Gilbert could not entirely ignore small variations over time (magnetic declination). The compass needle rarely pointed toward true north. In the northern Atlantic Ocean, the variation was always toward the nearer continent: near Europe, the variation was toward Europe; near America, it was toward America (fig. 2.3). If the Earth were a perfect sphere, Gilbert reasoned, the two directions would always coincide. However, if the Earth were not quite spherical, the unevenness of the Earth's surface could cause magnetic bodies to turn toward the more massive magnetic parts of the Earth.[2]

Throughout the first five books of *De Magnete*, Gilbert's experiments were conducted with miniature compass needles moved over the surface of a *terrela*. With these versoria, he managed to reproduce four of the five magnetic motions that he identified: the attraction of opposite poles (coition), north-south alignment (direction), variation, and magnetic dip. It was his success at reproducing these four motions with his apparatus that sustained his analogy that the Earth itself was a great magnet. Still, Gilbert was unable to reproduce the fifth motion—rotation on its axis—despite Peregrinus's claim to have shown that a spherical magnet suspended from its pole would rotate once every

24 hours. This failure did not dampen Gilbert's attempt to develop a magnetic cosmology of a moving Earth, one that was in step with the new heliocentric astronomy advanced by Nicolas Copernicus in *De Revolutionibus* (1543: *On the Revolutions*). It has been estimated that only 10 writers had voiced their support for the Copernican system by 1600. The most important of the early English Copernicans was Thomas Digges (c. 1546) who published his *A Perfit Description of the Caelestiall Orbes* in 1576, which translated key parts of Book I of Copernicus' treatise and presented the chief arguments in favor of the new system of Copernicus. He also removed the stars from a thin shell and extended them to infinity in all directions. Although Gilbert was not the first, he was certainly one of the most important of the first generation of Copernicans.

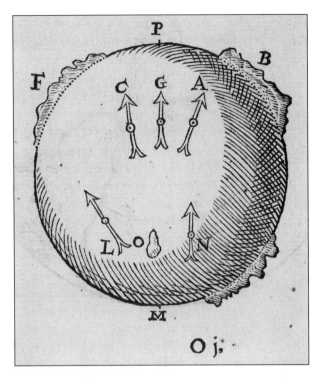

Figure 2.3: Gilbert's illustration of magnetic variation. From Gilbert, *De Magnete* (1600). With kind permission of the Thomas Fisher Rare Book Library, the University of Toronto.

The Earth's rotation is not occasioned by its shape, as Copernicus suggested, but by the magnetic and luminous power of the Sun, in concert with the Earth's own immaterial magnetic force; these magnetic forces combine to rotate the planet around its axis so that alternate faces are presented to the Sun and Moon in order to receive their virtues. The entire moving system of the planets is likewise ordered by mutual magnetic influences, which ensure the harmonious working of the whole. Magnetism, in short, causes the Earth's Copernican diurnal rotation. Perhaps in consequence of his having no experimental support for the annual revolution around the Sun, Gilbert makes no mention of the second and definitive motion attributed by Copernicus to the Earth, namely, its annual revolution around the Sun.

BACON AS A CRITIC OF GILBERT

One would be perfectly reasonable to suppose that Francis Bacon (1561–1626), long considered the leading exponent of experimental science during the first few decades of the seventeenth century, would have applauded the experimental bent of Gilbert's landmark book. It is therefore somewhat surprising that Bacon's assessment of Gilbert's work, published in his *Advancement of Learning* (1605), was anything but positive: "The alchemists have made a philosophy out of a few experiments of the furnace, and Gilbert, our countryman, hath made a philosophy out of observations of the lodestone" (cited by Benjamin 1895, 328). Despite Gilbert's being a practitioner of a brand of science openly lauded by Bacon, at every opportunity Bacon steadfastly opposed Gilbert's work, portraying it as an example of false scientific method; that is to say, as an error to be avoided.

When Bacon's position is fully taken into account, one can readily discern why he would take issue with Gilbert's work. "He [Gilbert]has himself become a magnet; that is, he has ascribed too many things to that force and built a ship out of a shell" (cited by Benjamin 1895, 328). It would not have been uncharitable for Bacon to have objected that Gilbert's strong thesis that the Earth's own magnetism caused its rotation went well beyond the mandate of the data. However, what Bacon refused to accept was Gilbert's weaker thesis that the Earth was fundamentally magnetic. By modern lights, Bacon's resistance to Gilbert's central claim is puzzling, especially when one is alert to the support that Bacon's thesis received from the discovery of the magnetic dip.

When one reflects on the manner in which the magnetic character of the Earth was understood by Gilbert, however, one can understand Bacon's reservations. For Gilbert, it is the soul or force that animates the Earth that serves to explain its Copernican motions. What's more, Gilbert went so far as to extend this animate force to the entire universe, and it is clear why Bacon would have taken exception to it. Gilbert's magnetic philosophy is shot through with Aristotelian convictions, the very convictions that Bacon and his fellow travelers were keen to discard. Gilbert had not articulated a science of magnetism but a magnetic natural philosophy, building, Bacon contends, a ship out of a shell. From Bacon's point of view, Gilbert's theory was essentially an immaterialist theory of properties of matter that were concealed to sensation and, in this respect, beyond the pale of science. Because Gilbert extended the Aristotelian model in many ways, his theory struck many readers as fundamentally Aristotelian in conception and, to this extent, out of step with the times.

Gilbert's magnetic philosophy did not take hold in England, in part because of the ruthlessness of Bacon's criticisms. On the continent, his work was received with more enthusiasm. Galileo (1564–1642), for instance, engaged in a long series of experiments in 1604 that are described in a series of letters composed during this period. In contrast to Bacon, Galileo was often effusive in his support for Gilbert's work: "From the work of *that grandissimo filosofo*, so exhaustive of its subject and so full of evident demonstrations, there can be no question that our earth in its primary and universal substance is none other than a great globe of lodestone" (cited by Fahie 1918: 246). Of course, we need to remind ourselves that Galileo's genius was to be felt in other fields of interest. As for magnetism, he was not interested in the phenomenon of magnetic induction and had no inkling of the mathematical possibilities of magnetic theory.

The result was that Gilbert's immaterial magnetic theory garnered little support. Indeed, as we shall see in the next section, one of the ironies of Gilbert's great work is that it stimulated interest in materialistic accounts of magnetism. Although his immaterial magnetic theory was stillborn, Gilbert's experimental work in electricity, which was based on a materialistic effluvial theory, was fruitful, signaling a return to the effluvial theory of Empedocles.

DESCARTES AND THE MECHANISM OF MATERIAL EFFLUVIA

One of the beneficiaries of the growing aversion to immaterialist accounts of magnetism was the mechanical theory of the French mathematician and philosopher René Descartes (1596–1650), whose purely mechanical theory of magnetism marks the pinnacle of magnetic science in the seventeenth century. Descartes insisted that all natural phenomena could profitably be reconceptualized as mechanical effects; that is, every natural phenomenon could be understood as effects of systems of interconnected mechanical devices. Following his mentor, the Dutch school teacher Isaac Beeckman (1588–1637), Descartes steadfastly opposed the notion of force, which he castigated as an occult concept that is either an unseen entity or a fictitious appearance. In either event, for Descartes force was fundamentally Aristotelian in conception.

Descartes founded his own alternative philosophy of nature on the assumption that apparent force can only be communicated by pressure or by the percussion of material bodies. This assumption required, in turn, that all of space must be filled with matter, not the ponderable matter of the Earth, but a subtle matter, or ether, that fills space everywhere. This ether pervaded all ordinary matter, as well as the space in the heavens. The circulation of the etherial medium accounted for the motion of the Sun, Moon, and planets, as well as the transmission of light and heat from the Sun and stars.

Descartes's magnetic theory is elaborated in his *Principia Philosophiae* (1644: *Principles of Philosophy*, Part IV, arts. 133–188). His theory is based on Empedocles's theory of effluvia. Long streams or emanations of tiny particles of matter, Descartes hypothesized, circulated through the pores of the magnet and iron (fig. 2.4), the movement of one particle followed by that of its neighbor, each taking the place of the other with no empty space between them. The guiding principle of Descartes's magnetic theory, then, is that attraction and repulsion are produced by the circulation of subtle magnetic matter in the space surrounding magnetic bodies, a position that contrasts with the immaterialist account of Gilbert that attributes attraction and repulsion to the bodies

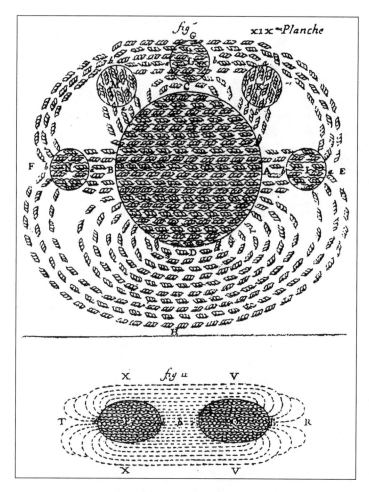

Figure 2.4: Corpuscular model of magnetic attraction. From Descartes, *Principia philosophiae* (1644).

themselves, with the action of matter in one body acting on the matter of the other body through intervening empty space. As we shall see, much of the opposition to Newton's theory of gravitation, interpreted as attraction at a distance across empty space, was sustained by the allure of an unmitigated mechanical cosmology.

The virtues of Descartes's magnetic theory were many, but chief among them was that one could frame an intelligible picture of the action of the magnet at the insensible level of minute particles of matter. Another virtue was that Descartes's theory explained why the iron became magnetized. When a piece of iron was subject to the flow of magnetic particles for an extended period of time, filaments in the iron would lose their elasticity and become oriented in one direction. The iron would thereby become magnetized and serve as a lodestone. As the projecting filaments slowly returned to their original position, the iron would lose its magnetic power. Consequently, a process that was used to make compass needles finally was given a theoretical explanation. According to Descartes, when a piece of iron was subjected to a flow of magnetic particles for a long period of time, the projecting filaments on the pores of the iron would lose their flexibility and become oriented in one direction. At this point, the iron would become magnetized.

The chief difficulty with Descartes's theory was that it was more of a general strategy for explaining magnetic phenomena than it was a theory in a more robust sense of the term. Indeed, Descartes's theory did not furnish a unified explanation for a cluster of magnetic phenomena but called for at least three distinct models—one to explain the orientation of the compass, another to explain the attraction and repulsion of different magnets, and a third to explain the attraction of the magnet for iron. Let's address these in order:

1. To explain the orientation of the compass toward north and south, Descartes hypothesized two streams of grooved particles, one with a right-hand groove and a second with a left-hand groove. The Earth was likewise fitted with pores that formed long passages oriented north-south, such that the grooved particles moved through the pores parallel to the axis of rotation. These pores were also grooved so that they would accept the circular streams of particles. The two circular streams circulated in two different directions, one north to south and the other from south to north. Particles that were intercepted by lodestone or magnetized iron exerted a pressure upon it, causing it to turn in the direction so that its grooved pores were aligned with the circulating streams.
2. To explain the attraction and repulsion of different magnets, Descartes submitted that the emanations of two magnets that were brought together with unlike poles facing each other were able to enter into the poles of the opposite magnet because the pores were disposed to receive the emanations. This mechanism of the flow of the circular vortex would force out the air between the magnetic poles. The air moving in a vortex in the manner of a circular thrust would move behind the magnets and drive them toward each other. When like poles faced one another, the magnetic particle flows from the magnets were unable to pass into the pores of the opposite magnet, resulting in a repulsive force because of the pressure created by the opposing streams of magnetic matter. To explain this, Descartes hypothesized that the pores were fitted with projections that acted like one-way valves, allowing grooved particles to pass in one direction and not the other. When the particles passed in one direction, the projections would as if flat, but they would ruffle up and impede the flow when they passed in the other direction.
3. To explain the attraction of magnet for iron, Descartes modified the mechanical arrangement used to explain the attraction of two different magnets. The pores of iron, Descartes claimed, contained projections arranged in a random fashion. When a lodestone was brought near the iron, the streams of grooved particles would enter the grooved pores of the iron and push their way along the channels because of the momentum of the magnetic flow. This, in turn, would cause the projecting filaments to be rearranged so that they would accept the direction of the magnetic flow. This process of magnetization would then be followed by the mechanism of attraction as the air was forced out from between the magnet and iron, and then moved around by the circular thrust.

Although all three models are based on the same explanatory strategy of swirling particles and appropriately situated pores, one wonders why there were different mechanisms for the various magnetic actions. One would presume that the same mechanism would work for all. What is especially glaring about Descartes's models is that they seem to be inconsistent with one another; that is, the mechanism for attraction and repulsion seem to rule each other out.

NOTES

1. Later, in his *Magnes Siue de Arte Magnetica* (1654: Magnesia or the Art of the Magnet), the Jesuit scientist Athanasius Kircher (1602–1680) objected that the Earth cannot be a large magnet simply because a magnet of that size would attract magnetic bodies with so great a force that they could not be prized off again, a possibility overlooked by Gilbert. Kircher's objection was anticipated by Descartes, who submitted that very small lodestones often have a force greater than the Earth because the Earth's magnetic virtue is concentrated in the bowels of the Earth and is feeble at its surface.

2. Based on the observation that the compass needle pointed west of the true north near the archipelago of Novya Zemlaya, in the Arctic Ocean north of Russia, Gilbert mused on the existence of a "north-east passage" around Russia, which, if discovered, would provide merchants more direct access by sea to the spice islands of the Far East. In fact, the Dutch explorer Willem Barentz (d. 1597), who was one of the most important of all Arctic explorers, had already made three voyages (1594, 1595, 1596–97) in search of the Northeast Passage to Asia.

3

ELECTRICAL CONDUCTION

BOYLE AND THE EXPERIMENTAL LIFE

Many of the leading intellectuals of the seventeenth century contributed to the elaboration of the burgeoning mechanical philosophy, which attempted to explain natural phenomena in terms of the collisions of minute particles of matter. However, Descartes was arguably the most vocal proponent of what has come to be known as "the mechanization of nature," as well as the architect of a comprehensive natural philosophy that was utterly uncompromising in character.

An enthusiastic proponent of the new mechanical philosophy was Robert Boyle (1627–1681), one of the leading scientific figures of the latter half of the seventeenth century. Where Descartes worked hard to defend the authority of reason, and was willing to consult experience only when his chain of reasoning broke down, Boyle was an assiduous experimenter, who was suspicious of reasoning in general and committed to the proposition that our theories should be grounded in and true to the facts as established by experiment. Although Francis Bacon was the most vocal proponent of a new science controlled by experiment, it was Boyle who realized such ideas in practice. Throughout his life, he displayed an extraordinary ingenuity in devising trials that would reveal significant information about the phenomena he studied, combined with an unprecedented precision in observing their outcome. One of the reasons for Boyle's success as an experimenter is the intellectual zeal with which he scrutinized the place of experiment in the context of scientific practice. His *Certain Physiological Essays*, which was published in 1661, mounted a compelling intellectual portrait of the value of experimentation, including one essay on the significance of failed experiments, and a second that detailed the manner in which experimental trial could be deployed to provide a foundation for the mechanical philosophy.

In 1656, Boyle had moved to Oxford, where he joined ranks with an active group of natural philosophers—a group, which he referred to as an "invisible college," that foreshadowed the founding of the Royal Society in 1660. At the same time, the young Robert Hooke (1635–1703) assumed a position as Boyle's assistant. Hooke played a significant role in the construction of Boyle's celebrated air pump (fig. 3.1), the idea of which was inspired by a demonstration performed two years prior by Otto von Guericke (1602-1686). With his pump, Boyle carried out a series of experiments, illustrating the characteristics and function of the air. Demonstrating that a vacuum could be produced with his air pump, he helped to discredit the ancient Aristotelian doctrine that nature abhors a vacuum These results were reported in Boyle's *New Experiments Physico-Mechanical, Touching the Spring of Air and its Effects* (1660).

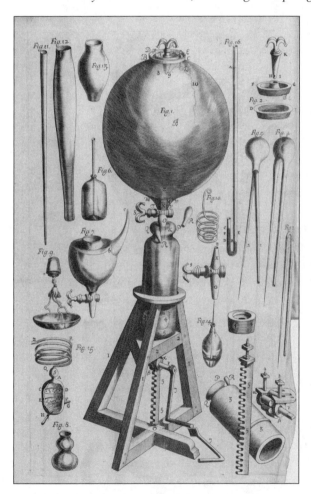

Figure 3.1: Engraving of Boyle's first air pump. From Boyle, *New Experiments Physico-Mechanical, Touching the Spring of Air and its Effects* (1660). With kind permission of the Thomas Fisher Rare Book Library, the University of Toronto.

Boyle published a steady stream of treatises on an impressive range of subjects throughout his life. One treatise that is particularly interesting, which has not received a great deal of attention from scholars of Boyle's works, is his *Experiments and Notes About the Mechanical Origine or Production of Electricity*. Published in 1675, this is the first book in English devoted exclusively to electrical phenomena. Once again, Boyle's penchant for careful and precise experiment disclosed a number of facts about electricity that were hitherto unknown. The most remarkable of these were that (1) amber retained its attractive power after the friction caused by the rubbing had ceased, (2) the attraction between two bodies was mutual, (3) heat and tersion (wiping of any body) increased its susceptibility to excitation, (4) a diamond in its rough state exceeded in its power of excitation all the polished bodies he had tried, thereby routing the ancient belief that smoothness of surface enhanced the excitation, and (5) a needle's response to a magnet did not change when it was placed inside a vessel evacuated of air. Boyle also added a number of substances to Gilbert's list of electrics, including white sapphire, white amethyst, a green stone thought to be a sapphire, and glass of lead.

As one might expect, in virtue of its importance to navigation, magnetism was studied extensively during the sixteenth and seventeenth centuries. Electricity, in contrast, was regarded during this period as a curious but comparatively uninteresting phenomenon. With the exception of Boyle's contributions, little electrical research was conducted before 1700, and the research that was carried out, including the work by Boyle, was in many respects a repetition of Gilbert's work. Aside from Gilbert's contributions to the subject, understanding of electricity had not advanced beyond what was known to scientists in ancient times.

GUERICKE'S ELECTROSTATIC APPARATUS

Another bright spot was the work of Otto von Guericke. Trained as an engineer, he built the first air pump in an attempt to create a vacuum and thereby refute the ancient belief that nature abhors a vacuum. At Ratisbon in 1654, he performed one of the most dramatic demonstration experiments in the annals of science, when, before the Imperial Diet, he showed how two teams of eight horses each could not separate a pair of hemispheres 12 feet in diameter from which he had exhausted the air with his air pump (fig. 3.2). When air was admitted, the hemispheres fell apart by themselves. Guericke also demonstrated that 20 men could not hold a piston in a cylinder once the air had been evacuated from one end of it. He demonstrated the air's weight and determined its density.

Figure 3.2: Guericke's successful demonstration of the pressure of air. Original Woodcut, ca. 1850. Courtesy of Corbis.

This dramatic demonstration of the pressure of air completely overshadowed Guericke's work in electricity, which was completed in 1663 but not published until 1672, with the appearance in print of his *Experimenta Nova* (*New Experiments*). In this work, Guericke describes a device that consisted of a sulfur ball revolving on an iron shaft. The ball was made by pouring molten sulfur into a spherical glass container "about the size of a child's head." When the sulfur cooled, the glass shell was broken away, leaving its spherical form. An iron shaft was inserted into the ball, and the ball and shaft were mounted onto a wooden frame. The ball was set in motion, and a dry hand was applied to it, thereby electrifying the sphere, which was then able to attract paper, feathers, and lint.

These light objects held fast to the rotating sphere, prompting Guericke to compare this phenomenon of electrical attraction to objects clinging to the surface of the Earth. In this way, he assigned an electrical cause to the attraction of objects to the Earth's surface, in contrast to Gilbert, who had suggested that this attraction was magnetic in nature. Guericke also noticed that drops of water brought near the electrified sphere became agitated, and that the attractive effect was dissipated when the sphere was brought near smoke or fire. Most important of all, he observed small sparks in the discharge and heard their crackling sound.

With his new device, Guericke observed the repulsion of what scientists would call *similarly electrified bodies*. He found that a glass rod that had been rubbed first attracted bits of paper to it and that, once they came in contact with the rod, would then be repelled by it. Electricity was imparted from the glass rod to the bits of paper, which were then repelled by it. Guericke also observed that a feather would move up and down between sphere and ground. Further observations showed that an electrical charge traveled out to the end of a linen thread (electrical conduction), and that bodies became charged even if only brought close to a charged sphere (electrical induction or influence). Guericke did not pursue these observations.

Guericke's sulfur globe proved to be surprisingly infertile. One reason is that the electrical effect was very difficult to reproduce; soft, moist hands all but ensured failure. Guericke traced his own aptitude to his former employment as a mechanic, but even he had difficulty making these experiments work. Alert to this problem, he distributed globes and provided lessons in their manufacture. Another consideration that may have dampened enthusiasm for Guericke's sulfur globe was that there was no other material, save perhaps for the Earth itself, which shared its properties. Because the property of the globe whereby it electrified bits of paper was not widely distributed in nature, this device globe could hardly be the foundation on which to found a science of electricity. The uniqueness of Guericke's sulfur globe stood in sharp contrast to Gilbert's efforts to demonstrate the generality of the amber effect.

NEWTON'S INTELLECTUAL CREED

Isaac Newton (1642–1727) is universally revered as the leading scientist of the early modern period (and arguably every period) in consequence of his monumental *Philosophia Naturalis Principia Mathematica* (1687: *Mathematical Principles of Natural Philosophy*), which laid the foundation for the reduction of phenomena to forces acting at a distance. In the "Preface" to the first edition of the *Principia*, we find the following celebrated statement of Newton's program:

> [T]he whole burden of philosophy seems to consist in this—from the phenomena of motions to investigate the forces of nature, and then from these forces to demonstrate the other phenomena ... I wish I could derive the rest of the phenomena of Nature by the same kind of reasoning from mechanical principles, for I am induced by many reasons to suspect that they may all depend upon certain forces by which the particles of bodies, by some causes hitherto unknown, are either mutually impelled towards one another, and cohere in regular forms, or are repelled and recede from one another. (Newton 1934, xvii.)

Newton's law of universal gravitation is expressed in the form

$$F_g = G \frac{m_1 m_2}{r^2},$$

where G is a constant, m_1 and m_2 are the masses of the attracting particles, r is the distance between them, and Fg is the force of gravity acting between the attracting particles m_1 and m_2. To say that a force is acting at a distance, then, is just to assert that its mathematical expression gives the force between two bodies directly in terms of certain properties of the particles themselves and of the distance between them (Woodruff 1962, 440). Such forces act instantaneously across the space between bodies. This approach contrasts sharply with the view, associated with the nineteenth century figures Michael Faraday and James Clerk Maxwell, which holds that the propagation of the influence of a particle takes time and occurs at a finite speed.

Newton's own view, however, did not support the idea of causally efficacious forces acting at a distance. Such a view, he proclaimed in a letter of 1693, was "so great an Absurdity, that I believe no Man who has in Philosophical Matters a competent Faculty of thinking, can ever fall into it" (Newton 1959–77, 3:253–54). What he proposed, rather, was to give the mathematical expression of a force as something acting at a distance—an enterprise that Newton regarded as quite distinct from a science that claimed to discern the cause of the force:

> I here use the word *attraction* for any endeavor whatever, made by bodies to approach to each other, whether that endeavor arise from the action of the bodies themselves, as tending to each other or agitating each other by spirits emitted; or whether it arises from the action of the ether or of the air, or of any medium whatever, whether corporeal or incorporeal, in any manner impelling bodies placed

therein towards each other. In the same general sense I use the word *impulse*, not defining in this treatise the species or physical qualities of forces, but investigating the quantities and mathematical proportions of them. (Newton 1934,192)

Although he pleaded agnosticism in the *Principia*—described by Newton as an exercise in experimental science—he was much more candid in Query 21 of his later *Opticks* (1704). Here he openly contemplates the possibility that an ether filling space was responsible for gravitational attraction, in addition to the role that it played in the propagation of light. Disturbances to this ether, he suggested, would not be propagated instantaneously.

ELECTRICITY AT THE ROYAL SOCIETY OF LONDON

Newton is frequently credited with the improvement of Guericke's electric apparatus by replacing the globe of sulfur with a globe of glass. There is no evidence for this contention, however, which appears to be based on a reference to a glass-globe generator in Query 8 of Newton's *Opticks* (1704).

Although Robert Boyle and other investigators of electrical phenomena regarded glass as a weak electric, Newton gave an experimental demonstration in 1675 of the usefulness of glass as an electric. He placed very small pieces of paper under a glass disk set in a brass ring so that the paper could not be moved by the motion of a draught of air between the glass and the table. He noticed that rubbing the glass on one side caused the paper to move on the other, that the pieces were attracted and moved to and fro for some time even after the rubbing had discontinued. Finally, he noticed that the motion of his finger on the upper surface of the glass would deflect the tiny fragments of paper that hung under the glass according to how he moved his finger. What this experiment demonstrated, Newton claimed, was that electrical effluvia can freely penetrate sensible thicknesses of glass (Newton 1959–77, 1:364). The Fellows of the Royal Society did their best to reproduce Newton's result but without any success.

Francis Hauksbee (1660–1713), an instrument maker, who was instructed in vacuum techniques (fig. 3.3) by his teacher Robert Boyle, entered the service of the society in 1703, the same day that Newton assumed the title of its president. The influence of Hauksbee on Newton is evident in the additions the latter made to the Queries in the second edition of the *Opticks* (1718) and in Newton's phrasing of the General Scholium, which he added to the second edition of the *Principia* (1713).

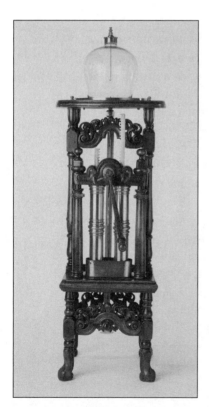

Figure 3.3: Hauksbee's air pump. Courtesy of Corbis.

The task that Newton assigned to Hauksbee was that of reviving the Royal Society tradition of showing experiments at its weekly meetings.[1] Hauksbee's initial interest was a phenomenon that first had been observed in 1675: When a barometer was carried at night, violent motion of the mercury would cause the appearance of a "fine purple light" in the empty glass tube just above the mercury column. This phenomenon of electrical luminescence was generally explained as due to the presence of phosphorus and sulfur in the mercury. Hauksbee's first appearance before the Royal Society in 1703 was with an air pump of his own design and experiments showing the "mercurial phosphor."

In October of 1705, Hauksbee returned to the phenomenon of the glow in the glass tube. He conjectured that the phenomenon was occasioned by the friction of mercury on the sides of the glass tube. In an attempt to generalize his conjecture to other materials, he built a machine for rotating amber on wool in a vessel evacuated of air (fig. 3.4). He was thereby able to observe the "induced" glow in the evacuated tube when a second, rubbed sphere was brought near it. Hauksbee did not draw any connection between the luminous phenomenon he had observed and electricity, so this discovery evaded his grasp.

EARLY EXPERIMENTS IN ELECTRICAL CONDUCTION

One of the most ardent experimenters of the early eighteenth century was the Englishman Stephen Gray (1666–1736). After some basic schooling, Gray was apprenticed to his father (and subsequently his elder brother) in the cloth-dyeing trade. His interests, however, lay with natural science and particularly with astronomy, and he managed to educate himself in these developing disciplines, mainly through wealthy friends in the district who permitted him to access their libraries and scientific instruments.

Gray prepared his own lenses and constructed a telescope, with which he made a number of minor discoveries, principally in the study of sunspots. These discoveries, some of which were published by the Royal Society, secured Gray's reputation as an astute observer

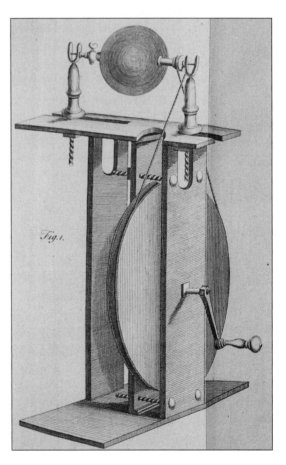

Figure 3.4: Hauksbee's device for rotating amber in a vessel evacuated of air. From Hauksbee, *Physico-Mechanical Experiments on Various Subjects.* (1709).

of natural phenomena. Publication attracted the attention of John Flamsteed (1646–1719), the first English Astronomer Royal, who was engaged in constructing the new observatory at Greenwich. Flamsteed's astronomical activities were geared toward the creation of the first detailed and accurate star catalogue.

Gray assisted Flamsteed with many of his observations and calculations. This entrance into the world of science was not without a cost. Flamsteed was embroiled in a protracted dispute with Isaac Newton, who presided over British science with an iron fist, over access to preliminary star-chart data. This dispute erupted into a full blown war in the Royal Society, which was controlled by Newton, ensuring the marginalization of Flamsteed and his associates for decades (for a thorough discussion of this controversy, see Clark 2001).

Newton, who was anxious to publish a second edition of his *Principia*, was certain that Flamsteed's data would provide support for his lunar theory. He was also convinced that Flamsteed was deliberately withholding information from him. But Flamsteed, who was every bit the perfectionist that Newton was, insisted that the calculations required to make accurate star catalogues did not come easily—and he had 30,000 figures to check for accuracy (at the rate of two hours per figure), for a catalogue that would eventually list 3,000 stars. As a sign of good faith that he was working on perfecting the definitive catalogue, Flamsteed provided Newton with a sealed copy of his "draft" version of the star catalogue, which Flamsteed knew contained many inaccuracies.

Newton, not surprisingly, broke the seal and immediately sent the work to a printer, hiring his friend Edmund Halley (of Halley's Comet fame) to proofread the galleys and correct the erroneous figures. Halley detected something was afoot and let it be generally known that Newton was about to publish an illicit, inaccurate, and pirated version of Flamsteed's life's work (Clark 2001). Halley even offered to publish the definitive work at his own expense, in an attempt to restrain Newton. The pirated version of Flamsteed's work—*Historia Coelestis Britannica*—was published in 1712, just the same.

After the fray between Newton and Flamsteed had begun to widen, Newton was made head of the British Mint and, later, president of the Royal Society. With the considerable power these two offices conferred on him, Newton proposed building an observatory at Trinity College to rival the one at Greenwich, and thereby bypass his reliance on Flamsteed's astronomical results.

Around this time, Gray was given free copies of the Royal Society's official publication, *Philosophical Transactions*, by Hans Sloan (1660–1753), secretary to the Royal Society. Between 1706 and 1711, the *Philosophical Transactions* published no less than 10 of Gray's reports of his experimental work in electricity. After Newton became president of the Royal Society, however, and until his death in 1727, Gray's gratis subscription to the *Philosophical*

Transactions ended, and the journal published only one article by him. However, after Newton's death and upon Hans Sloan taking over the presidency of the Royal Society (1727–1741), the *Transactions* once again regularly published Gray's letters and resumed his complimentary subscription.

During these years, Hans Sloan had kept copies of all the letters that Gray had sent to the Royal Society. Sometime after his death, these letters were discovered in Sloan's archives. They testify that Gray had continued to submit letters to the society during the years he was not being published. The letters testify to the innovativeness of Gray's experiments in electricity, and the degree to which Newton's comrade in arms, Francis Hauksbee, who was given access to Gray's unpublished letters, appropriated Gray's ideas in the articles that he, Hauksbee, published in the *Transactions*.

Through the efforts of Flamsteed and Sloan, Gray managed to obtain a pensioned position at the Charterhouse in London, a home for impoverished gentlemen who had served their country. During this time, he returned to his experimenting with static electricity, using a glass tube as a friction generator. During the course of one experiment in 1729, he noticed that the cork, which had been placed in the end of his glass tube as a barrier to moisture and dust, attracted small pieces of paper and chaff when it was rubbed. He tried extending the cork with a small fir stick plugged into the middle, with the result that the charge was evident at the end of the stick. He then tried longer sticks, with the same result, so he added a length of packthread—a two- or three-ply twine used for sewing or tying bundles—wound around an ivory ball. The result was the same: The ivory ball attracted light objects as though it were the electrified glass tube. The electric virtue, Gray observed, was carried over distance—certain bodies possessed the power of conveying electricity from one body to another or, in modern terminology, of *conducting* it. This was a result of enormous importance. Prior to Gray's discovery, the only way of producing the attractive emanations was by friction. Over the course of a few days, Gray managed to transmit electricity from a piece of rubbed glass through packthread supported by silken cords a distance of 765 feet, and through wire at a distance of some 850 feet.

In the course of these experiments on electrical conduction, Gray discovered the importance of insulating his thread wire from ground contact. The experiment failed when the thread was suspended horizontally with strings made of hemp. Silk, which was drier and so less a conductor than hemp, did not leak away the electrical charge and so proved more successful. Gray also discovered that electricity was carried around bends in the thread and that, when the thread was dropped from a tower, it was not assisted by gravity.

From these experiments came an understanding of the role played by conductors and insulators—names coined by John Desauliers (1683–1744), demonstrator of experiments for the Royal Society and inventor of the planetarium. Desauliers repeated many of Gray's experiments and found, in addition, that bodies that possess the property of being electrically excitable by friction, or *electrics per se,* do not have the power of conduction, whereas conductors are not electrics per se.

Gray also experimented on electrical induction or electrification by influence. Having suspended a lump of lead from the ceiling by a thread, he brought a rubbed glass tune near it and found that the lead attracted and repelled brass filings. He concluded that the electrical virtue can be transmitted without contact.

When Sloan took over the Royal Society on Newton's death, Gray belatedly received the recognition denied him by Newton. He was given the Society's first Copley Medal in 1731 for his work on conduction and insulation, and also its second Medal in 1732 for his induction experiments. In 1732 he was also admitted as a member of the Royal Society, but he died destitute a few years later in 1736.

NOTE

1. Hauksbee's many articles published between 1703 and 1709 were gathered together and published under the title *Physico-Mechanical Experiments on Various Subjects.* Articles published after 1709 and until his death in 1713 were published in a second, expanded edition that appeared in 1716.

4

ELECTROSTATIC PHENOMENA

Early researches in electrical conduction were limited by the quantity of electricity generated by frictional means. A giant step was taken by science with the discovery of a means of accumulating and preserving electricity in large quantities. The principle of the *condenser* —a device that stores energy in an electric field by accumulating an internal imbalance of electric charge—was discovered independently in 1745 by Pieter van Musschenbroek (1692–1761) and Ewald G. von Kleist (1700–1748). Historians have not reached a consensus regarding who should be credited with this discovery. Von Kleist was arguably the first to discover the principle of the condenser, but it was Musschenbroek who announced this discovery in a manner that enabled scientists to reproduce his results for themselves. Musschenbroek carried out his work at Leyden University in the Netherlands. In recognition of his role in bringing this discovery to the scientific community, the original form of the condenser came to be known as the Leyden jar.

Musschenbroek was born into a family of instrument makers. After completing a degree in natural philosophy at the University of Leyden (Leiden), he visited England in 1717 and met Isaac Newton. Upon returning to the Netherlands, he held professorships (from 1721) at the universities of Duesberg (Duisburg), Utrecht, and Leyden (Leyden from 1740 to 1761). In addition to his contributions to the science of electricity, he is remembered for introducing Newton's ideas to the Netherlands.

As a follower of Newton, Musschenbroek openly declared his opposition to the Cartesian theory of circulating magnetic matter in a letter published in the *Philosophical Transactions of the Royal Society* in 1725. Experiments, he insisted, demonstrated the falsity of the Cartesian theory. This case against Descartes was elaborated in a treatise on magnetism published in 1729.

Figure 4.1: The Leyden experiment. From Nollet, *Essai sur l'éléctricité des corps* (1746).

The most telling experiment was one that claimed to show that the attractive strengths of the north and south magnetic poles were different. If the magnetic flow behaved according to the known laws of fluid flow (i.e., if it behaved as a fluid vortex of matter), the flow at the opposite poles was required to be the same. Because the flow was not the same, Musschenbroek concluded that Descartes's theory was mistaken. But it was Musschenbroek's fact-finding that was in error. The erroneous belief that the magnetic flow was not the same strength at opposite poles was taken as a refutation of the Cartesian theory.

In addition, there was the old argument that material bodies interposed between magnets did not change the magnetic forces and that only iron bodies had this effect. This result could not be explained by the Cartesian theory that relied upon the passage of magnetic particles to account for its magnetization, yet these particles were completely unaffected by their travel through bodies other than iron or steel.

In 1745, Musschenbroek and a small group of collaborators created a novel device in an attempt to prevent the leaking away of electrical charge, which he conjectured was occasioned by air conduction. Musschenbroek's innovative idea was to prevent leakage by surrounding the charged body by a nonconductor. And so, he suspended a bottle of water from a gun barrel by means of a metal wire passing through the cork (fig. 4.1). The gun barrel was hung up by silk threads and charged by means of an electrified body brought near. What he expected was that this charge would pass into the water, the glass bottle forming the nonconductor that would protect the charge. He was therefore startled when he touched the gun barrel with one hand and the glass bottle with the other. The jar device had accumulated the electricity produced by the static machine, and then all at once it discharged to Musschenbroek. The effect made a vivid impression on him, as conveyed by the following excerpt from one of a series of letters that he sent to the Académié Royale des Sciences in 1746:

I wish to advise you of a new but terrible experiment, which I advise you on no account personally to attempt. I am engaged in a research to determine the strength of electricity. With this object I had suspended by two blue silk threads, a gun barrel, which received electricity from communication by a glass globe which was turned rapidly on its axis by one operator, while another pressed his hands against it. From the opposite end of the gun barrel hung a brass wire, the end of which entered a glass jar, which was partly full of water. This jar I held in my right hand, while with my left I attempted to draw sparks from the gun barrel. Suddenly I received in my right hand a shock of such violence that my whole body was shaken as by a lightening stroke. The vessel, although of glass, was not broken, nor was the hand displaced by the commotion: but the arm and body were affected in a manner more terrible than I can express. In a word, I believed that I was done for. (cited in Benjamin 1895, 517)

A means of accumulating the electric force had been created. In its improved form, the Leyden jar consisted of a glass jar—the thinner the glass, the more effective was the discharge—whose outer and inner surfaces were covered with tinfoil. A brass knob was fixed on the end of a stout brass wire, which was then passed downward through a lid of dry, well-varnished wood. The jar was half-filled with water and the wire dipped into it, making the jar capable of storing an electric charge. To increase the charge, several jars could be arranged in a metal pan, thus combining their effects.

A charge was communicated to the condenser by holding it near an electric machine. When a positive electric charge was imparted to the inner coating of the jar, it acted inductively on the outer coating, attracting a negative electric charge to the face of the outer coating nearest the glass. If the jar was dry, free of dust, and made of good glass, it would retain its charge for days. The condenser would be discharged if one hand came in contact with the outer coating of the jar while the other hand simultaneously touched the knob on the lid. A bright spark would pass between the knob and the hand with a sharp retort and an ensuing shock to the muscles of the wrists, elbows, and shoulders. To avoid injury, discharging tongs, consisting of a jointed brass rod provided with brass knobs and a glass handle, were developed. One knob was laid against the outer coating, and the other was then made to touch the knob of the jar.

Jean-Antoine Abbe Nollet (1700–1770), the first professor of experimental physics at the University of Paris and prominent member of the French Academy of Sciences, was the first French scientist to experiment with the Leyden jar, along the way contributing greatly to its improvement. He found that other liquids, such as mercury, served just as well as water as repositories for the electric charge. He tested the effects of electricity on plants and animals (fig. 4.2). In one experiment, he planted mustard seeds in separate receptacles and kept one of them under an electric charge for eight days. He claimed that the electrified plants grew more than four times as fast as the others. To study the effects of electricity on animals, Nollet selected two cats, each four months old. One cat was kept near an electrical machine and was given

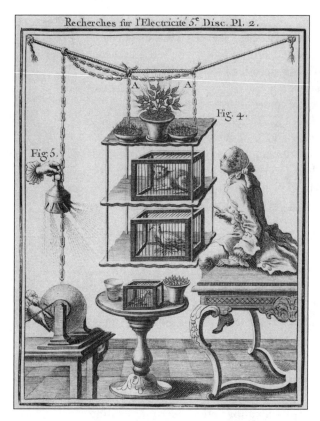

Figure 4.2: Experiments with a Leyden jar on animals and plants. From Nollet, *Recherches sur les causes particulieres* (1749).

electric treatments. This cat lost more weight than the other. The experiments were extended to small birds, to pigeons and to humans, with consistent loss of weight. He then tried to administer electric treatment to sick patients, with no positive results.

As one can well imagine, this device made for exciting and sometimes gruesome electrical demonstrations. Small birds and mammals could be killed with discharges from Leyden jars. The first wave of electricians who tried the Leyden jar reported numerous side effects: fevers, temporary paralysis, concussions, convulsions, and dizziness, among others. Despite these reports and Musschenbroek's theatrics, none of the early researchers were martyred by the Leyden jar. All were expressing their uncertainty in the face of a spectacular new source of the electrical force that greatly surpassed anyone's expectations. All were equally frank in their admission that the new condenser violated received principles of electricity and that they now understood nothing.

Experiments on electrical conduction took on a new dimension with the accumulation of ample amounts of electricity. The Royal Society, with William Watson (1715–1787) as the chief operator of its electrical machine, carried out a series of experiments in August 1748 with the purpose of determining the velocity of the electric fluid. At the time, it was widely believed that the velocity of electricity exceeded that of sound, but no accurate test had been devised to measure the velocity of a current. In fields north of London, Watson laid out a line of wire supported by dry sticks and silk which ran for 12,276 feet. He found that, even at this length, the time occupied in the passage of electricity was altogether inappreciable.

IMPROVEMENTS TO INSTRUMENTATION

Gilbert's crude, qualitative, pivoted electroscope, or versorium, was the earliest detector of the electrical charge. More than a century passed before an effort was made to improve this device. In 1731, Stephen Gray introduced a simple hanging thread—called a "Pendulous thread"—that would be attracted to any electrified body nearby. John Canton (1712–1782), who is remembered for his contributions to electrostatics, as well as for being the first British scientist to verify Franklin's conclusion regarding the identity of electricity and lightning (see the discussion in chapter 5), greatly improved this device by adding two small pith balls suspended by fine linen thread. The upper ends of the threads were fastened inside a wooden box. When placed in the presence of a charged body, the two balls would become similarly charged, and because like charges repel, the balls would separate. The degree of separation was a rough indicator of the amount of charge.

In 1770, Tiberius Cavallo (1749–1809), who designed numerous electrostatic machines and was responsible for the rediscovery of Peregrinus's long-lost letter on the magnet, made an electroscope of silver wires with pith balls at their ends, which was enclosed in a glass jar with tinfoil strips fixed down the sides.

Quantitative measurement was introduced the same year by the English electrician William Henley (c. eighteenth century), who hung a pith ball on a thin piece of cane pivoted on top of a quadrant. A charge brought near the ball caused the arm to move angularly outward, indicating the degree of motion on a graduated ivory semicircular scale fixed to the quadrant. With this *electrometer*, Henley compared the conducting powers of different metals by the charges required to melt one inch of wire of each metal.

The first true electrometer was the work of Horace Benedict de Saussure (1740–1799), renowned Alpine explorer and professor at the University of Geneva, who placed the strings and balls inside an inverted glass jar and added a printed scale so that the distance or angle between the balls could be measured. De Saussure made the important discovery that the distance between the balls was not linearly related to the amount of charge, but he did not follow up on this vital clue to the relationship between distance and charge. The exact relationship would not be ascertained until 1784 with the work of Charles Coulomb (see the discussion in chapter 5).

The physicist Abraham Bennet (1750–1799) invented the gold-leaf electrometer that is still in use today. In 1787, he arranged a glass bell jar capped by a brass disk from which a short brass rod or tube projected down into the jar. Two strips of gold leaf were attached to this rod, and additional strips of tinfoil were pasted on the inside of the jar facing the gold-leaf strips. The jar eliminated the influence of moving air. The divergence of the strips in Bennet's electrometer was measured on a graduated paper scale. The instrument was sufficiently sensitive to register the evaporation of water from a metal vessel placed on the cap of the jar.

Alessandro Volta (1745–1827), whom we will have occasion in a later chapter to discuss in greater detail, secured his reputation in 1775 with the announcement of his famous *electrophorus,* or carrier of electricity, which

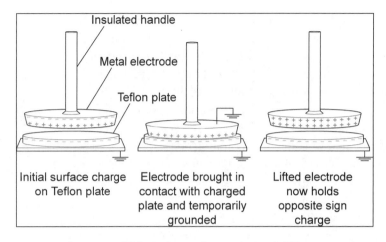

Figure 4.3: Diagram of Volta's electrophorus. Artist: Jeff Dixon.

worked by electrostatic induction, rather than by the then well-known direct electrostatic frictional means. His device consisted of a cake or resin, wax, or other non-conducting substance placed between two metal disks.

The resin rested on the lower disk, and the upper disk had an insulated handle attached to its center that permitted it to be lifted from the resin cake. To use the apparatus, the upper metal plate was removed, and the upper surface of the resin was rubbed or struck by a piece of flannel or animal fur, which thus charged the resin cake. The upper plate, held by the insulated handle, was then placed on the resin cake, and by touching the top plate with a finger, the charge was drawn off to ground. On lifting the upper plate, it would thus be charged by induction. A spark could then be drawn from the plate, and when discharged, the plate could once more be placed on the cake and recharged, and then again recharged. This repetition is what gave this carrier of electricity its name, *elettroforo perpetuo*. Although the idea of induced charges had been known for many years, no earlier apparatus had been constructed on this principle. So popular did this device become that many were built and some were made as large as seven feet in diameter.

5

FROM EFFLUVIA TO FLUIDS

ELECTROSTATIC ATTRACTION AND REPULSION

Gray's experiments marked an important transition in the history of the theory of electrical effluvia. After Gray's discovery of electrical conduction, it was no longer possible to believe that the electrical effluvia were inseparably connected with the bodies from which they are evoked by rubbing. Indeed, because these emanations could be transferred from one body to another, it appeared to scientists that these emanations had a separate existence—that they were an elementary substance, on a par with light, caloric, and other so-called "imponderable substances" that were thought to be destitute of weight.

The suggestion that the electric fluid was an element was opposed by many scientists, who were convinced that electricity should be explained in the same way as heat. Both heat and electricity could be induced by friction, both could induce combustion, and both could be transferred from one body to another by mere contact. Furthermore, the very best conductors of heat were also the best conductors of electricity, cementing the perceived analogy between electricity and heat.

This position was countered by the obvious knowledge that the electrification of a body did not produce any increase in its temperature. What's more, Stephen Gray conducted an experiment in 1729 involving the electrification of two cubes made of oak, one that was solid and the other hollow. This experiment, which was printed in the *Philosophical Transactions* (vol. 12) in 1739, indicated that, when these bodies were electrified in the same way, they produced the same effects. Because these objects were made of the same material but differently shaped, he concluded that it was only the surface of the cubes that was electrified. Whereas heat was distributed throughout the entire body, the electric fluid appeared to reside at or near its surface.

Charles François de Cisternay Dufay (1698–1739), a chemist by trade who is now remembered for a number of memoirs on electricity published between 1733 and 1737, visited Gray in 1732 and witnessed his experiments on conduction. Returning to France, he repeated Gray's experiments but this time he soaked the packthread in water and conveyed electricity some 1256 feet.

Although he was primarily an experimenter, Dufay formulated the first general theory of electricity called the *two-fluid theory*, according to which there were two kinds of electricity with different mechanical properties that attracted one another while each repelled itself. The electricity produced on glass by rubbing it with silk was called "vitreous" electricity (from the Latin term for "glass"); and the electricity excited in resinous bodies, such as sealing wax, amber, paper, and silk thread, was called "resinous" electricity. Bodies with vitreous electricity attracted all bodies with resinous electricity; and repelled all bodies with vitreous electricity. Two electrified silk threads, experiments showed, repelled one another. To ascertain whether a body was vitreous or resinous, one merely had to see whether it attracted an electrified silk thread. If such a thread was attracted by the body, it was vitreous; if the silk thread was repelled, the body was resinous.

Dufay's two-fluid theory of electricity was taken up by Nollet, and as we shall see in the next section of this chapter, opposed (to a degree) by Benjamin Franklin (1706–1790) in Philadelphia by a single-fluid, two-state theory of electricity. The two-fluid theory enjoyed considerable support, in part because of the enormous contributions that Dufay made to the subject of electricity. His memoirs testify that he attempted to electrify every kind of substance and object that he could lay his hands on. In the process, he showed that all bodies, with the exception of metals and those gums that softened on being chafed could be electrified. Consequently, what was once thought to be the special property of amber and a few other substances was shown by Dufay to be a property that was widely distributed throughout nature. In this manner, Gilbert's classification of bodies into "electrics" (capable of being electrified by friction) and "non-electrics" (not possessing this property) was shown by Dufay to be without merit.

Dufay also worked on the practical problem of insulating wires from electrical discharge. He tried all kinds of materials and objects as supports to the wires, and also varied the nature of the electrified object to which the wires were attached. He found that glass was one of the best means of isolating the object so that it did not lose its charge. He also made the startling discovery that the metals, formerly so poor at holding charge, could in fact hold the strongest charge if they were properly insulated.

ATTRACTION AND REPULSION

The invention of the Leyden jar generated enormous interest in the study of electricity among European scientists who routinely exchanged information on important new experimental developments. During the lifetime of Newton, electricity had attracted little attention, but by the middle of the eighteenth century, interest in the study of electrical phenomena had assumed international proportions with the participation of the American statesman and inventor Benjamin Franklin.

It is remarkable that Franklin emerged as one of the dominant figures in electrical research. Although there can be no doubt that Franklin's isolation was to some extent beneficial because it freed his work from some of the prejudices that held his European counterparts in their grip, it is equally true that he was frequently ignorant of the latest empirical research.

Franklin was in many ways representative of the American ideal. His father had left England for religious reasons and, in New England, had married and fathered 14 children. Benjamin, who had shown remarkable learning ability, was apprenticed to his older brother, who was a printer by trade. Striking out on his own as a typographer and bookseller, Franklin soon moved to London where he aspired to meet Newton through an acquaintance with Henry Pemberton (1694–1771), editor of the third edition of Newton's celebrated *Principia* (1726). This meeting did not occur, but Franklin did meet Peter Collinson (c. 1693–1768), a Fellow of the Royal Society and an important figure in his later electrical researches.

On returning to Philadelphia in 1726, Franklin branched out into journalism, publishing the celebrated *Poor Richard's Almanac*. With this publication, his fame started to spread and he received official appointments and became influential in Pennsylvania politics. He also amassed a fortune, mostly due to his penchant for practical inventions, such as the famous Franklin stove, which improved fuel economy.

At the comparatively advanced age of 40, Franklin was drawn to the subject of electricity by a series of electrical experiments that he witnessed by chance on the occasion of a visit to Boston. From the beginning, his electrical intuitions were shaped by the violent sparks of the Leyden jar, in contrast to scientists across the Atlantic who had been working with electrified bodies for some time and for whom electricity was synonymous with its powers of attraction and repulsion—powers that had stood front and center in the study of electricity since antiquity. For these scientists, the science of electricity had to be true to these phenomena, and this presumption, in turn, added a great deal of credibility to the theory of an electrical effluvium that somehow swirled out from a rubbed body and impinged upon the motion of light objects.

Although the effluvium theory furnished a compelling explanation for the phenomena of attraction and repulsion, it gave an unsatisfactory explanation for how effluvia stored in the jar acted on bodies situated at a distance from the jar. Nollet, for example, suggested that the effluvia accumulated on one side of the jar and, upon discharge, were released and somehow managed to penetrate through the glass—an account that was manifestly unsatisfactory. Franklin

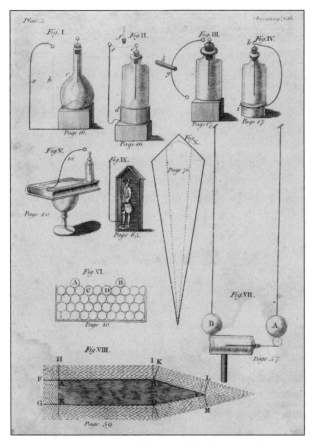

Figure 5.1: Electrical apparatus. From Benjamin Franklin, *Experiments and Observations on Electricity* (1751).

simply ignored this problem, feeling no need to explain the phenomena of attraction and repulsion.

Franklin worked with three assistants—Philip Syng (1703–1789), a skilled silversmith; Thomas Hopkinson (1709–1751), president of the American Philosophical Society; and Ebenezer Kinnersley (1711–1778), a Baptist minister without a church, who became a major spokesman for Franklin's discoveries. The first order of business of this working group was setting up a crank-turned machine to generate electric charge (fig. 5.1). By early 1747, they had acquired a Leyden jar to collect the charge. Throughout this year, Franklin sent a number of letters, reporting his discoveries to Collinson, the first of which involved the shape of conductors. Franklin found that a blunt conductor, when grounded, could draw off a charge from a distance of about an inch, and then only through a spark. A pointed conductor, however, could draw off a charge from a distance of six to eight inches, with no spark.

In short order, Franklin elaborated his own rival theory of electricity, according to which there was only one kind of electricity. During the course of experiment, Franklin had found that when a piece of glass is rubbed with a cloth, the glass receives a charge of the same strength as the charge on the cloth but opposite in kind. On this basis, he submitted that all bodies possess a natural quantity of a single fluid or "electrical fire" that is in a neutral state of electrical equilibrium until, through friction, an electric charge develops. This charge, in Franklin's theory, is an imbalance in an otherwise normally neutral state. If a body loses some of this fluid when rubbed, Franklin called it electrically negative, or minus; if it gained some, he called it electrically positive, or plus. These phenomena are now known to be due to an excess or deficit of electrons on the material that is rubbed; that is, materials that acquire electrons when rubbed gain a net negative charge; those that lose electrons gain a net positive charge. By modern lights, Franklin's convention was backwards.

Franklin's one-fluid theory of electricity helped to explain many of the subtle results that were observed when different kinds of substances were rubbed by various materials. A positive charge corresponded to the "vitreous" electric fluid, and a negative charge to the "resinous" fluid. Franklin thus proposed two conditions of a single fluid to replace a system of two different fluids. Electricity was really "vitreous" electricity and positive; its absence or depletion was "resinous" and negative. He also proposed the simpler and more meaningful terms "conductor" and "non-conductor" to replace the older "electric per se" and "non-electric."

Franklin demonstrated his theory concerning a single kind of electricity by having two people stand on insulated platforms made of wax, one of whom rubbed a glass tube with a cloth and took the charge from the cloth, while the other took the charge from the glass. When they brought their fingers close together a strong spark passed between them and both were completely discharged, showing that the charges had neutralized each other. Another demonstration of the same principle was made by hanging a cork ball between two knobs connected to the inner and outer linings of a Leyden jar. The ball would vibrate between the two knobs until the Leyden jar was completely discharged and both coatings were neutral with respect to the Earth.

Franklin's one-fluid theory offered a straightforward explanation of electrostatic induction. If a positively charged object was brought near an uncharged body, the excess of fluid in the first would repel the fluid in the second and drive it toward the farthest portion of the uncharged body, leaving the nearest portion of that body negatively charged and the farthest portion positively charged. There would now be an attraction between the positively charged body and the negatively charged portion of the uncharged body. However, since the positively charged portion of the uncharged body was farther from the positively charged body than was the negatively charged portion, the force of repulsion would be weaker than the force of attraction, resulting in a net attractive force. The same approach was used to explain what happens when a negatively charged body approaches an uncharged body.

The guiding principle of Franklin's theory was that electric charge can neither be created nor destroyed. Franklin was not the first scientist to identify this principle of the conservation of electric charge, but he was the first to make it the cornerstone of a theory of electricity. If a positive charge is produced by the influx of electric fluid, the implication was that this fluid must have come from somewhere else, and a deficit must exist at the point from which it came. The deficit produced at its point of origin must exactly equal the excess produced at its point of final rest. Accordingly, if glass is rubbed with silk and if glass gains a positive charge, the silk gains an equal negative charge. The net electric charge in glass-plus-silk was zero before rubbing and after.

Franklin's theory suffered from one drawback. It could not explain the repulsion of two negatively charged bodies; why a mutual lack of electrical fluid should have this effect was a mystery. There were only two ways out of this difficulty. Either the particles of matter should be considered as having repulsive as well as attractive virtues, or two electrical fluids needed to be assumed. The first option was advanced in 1759 by the German scientist Franz Aepinus (1724–1802) in his *Tentamen theoriae electricitatis et magnetismi (An Attempt at a Theory of Electricity and Magnetism)*—a book that marks the first attempt to apply mathematics to the theory of electricity and magnetism. The second was suggested by Charles Coulomb in 1788. Both approaches would have supporters during the nineteenth century.

Franklin brought this theoretical position to bear on the theory of the Leyden jar. Experiment convinced him that the charge resided in the *glass* of the jar, not in the outer foil or in the inner liquid, as was widely believed to be the case. To bolster this thesis, he set his Leyden jar on a glass insulator. After the cork and wire were removed, he found that the jar could still be discharged if he touched the outside and the water. He then moved the fluid from a charged jar to an empty uncharged jar resting on glass. This second jar would not discharge. However, when he refilled the first jar with pure water, he found that it had retained its power to shock. The evidence pointed to the glass as the source of the shock.

Franklin termed the state of electrical charge on the outside of the jar positive and that on the inside negative, and decided that rubbing simply caused the transfer of the charge from one surface to the other, the gain and the loss being of exactly equal quantity. He demonstrated this contention by attaching a wire to the outer coating of a jar and passing another wire through the cork and into the water on the inside. Between the terminals of these wires, he suspended a cork ball by a silk thread. He then placed the jar on a cake of wax to prevent leakage of the charge to the table or the floor. On electrifying the jar, the cork would oscillate from one terminal to the other, "fetching fire" from the inside until the equilibrium was restored and the cork ball hung limp between the wires.

Franklin brought these insights to bear on similarities noted by Nollet and Gray between lightning and the electric spark, and between the artificial snap of electricity and natural thunder. The prevailing opinion was that thunder and lightning were due to exploding gases. Some believed the phenomena were due to the detonation of a mixture of nitrous and sulfurous vapors in the air—the conditions being similar to those occurring during the explosion with gunpowder—although opinion differed as to the nature of the gases.

In a letter to Collinson in 1749, Franklin had written:

> The electrical spark is zigzag, and not straight; so is lightening. Pointed bodies attract electricity; lightning strikes mountains, trees, spires, masts and chimneys. When different paths are offered to the escape of electricity, it chooses the best conductor; so does lightning. Electricity fires combustibles; so does lightning. Electricity fuses metals; so does lightning. Lightning rends bad conductors when it strikes them; so does electricity when rendered sufficiently strong. Lightning reverses the poles of a magnet; electricity has the same effect (qtd. in Motteley 1922, 194).

At about the same time, Franklin had published his plan for an experiment to ascertain from elevated structures whether the clouds that contain lightning are electrified or not. He first proposed to use a tall spire to be constructed in Philadelphia to "draw off" electricity from thunderclouds overhead, but delay in its construction scuttled this plan. Following this suggestion, in 1752 two French researchers set up a tall, insulated pole in the village of Marly-la-Ville. Considerable time passed, and the researchers returned to Paris by the time the storm finally materialized. It was left for two townsfolk, who served as their assistants, to draw sparks by touching the pole with a brass wire mounted in a glass handle.

To test his theory that the Leyden jar produced miniature bolts of lightning and tiny peals of thunder, Franklin constructed a kite out of light strips of cedar and a large handkerchief made out of fine silk. A metal rod was then attached to the wooden framework of the kite. A tail and a length of twine completed the assembly. A silk ribbon was affixed to the end of the twine next to where his hand would be positioned. An iron key, attached to the junction of the silk and twine, completed his apparatus. This he raised, in company with his son, into a thunderstorm in the month of June 1752. With this apparatus, he proposed to bring the fires of the heavens down to the Earth, so that they could be placed side by side the sparks and flashes of his jars and their identity would be evident to all.

The Franklins positioned themselves inside a door, as a strategy for keeping the silk ribbon dry. As the kite vanished into one of the clouds, Franklin senior noted the fibers of the kite cord standing apart as though all were charged and repelling each other. The key, it appeared, had gained a charge. Cautiously, he brought the knuckle of his hand near the key. As he did so, a spark leapt out— the same kind of spark along with the characteristic crackle of the discharge of a Leyden jar. Franklin then brought out a Leyden jar that he had brought with him and charged it with electricity from the clouds. He thereby conclusively demonstrated that there was electricity in the clouds above, just as there was on the ground; that lightning was a gigantic electrical discharge; and that thunder was the resounding crackle that accompanied it. The chemist Joseph Priestley (1733–1804), who was a friend of Franklin's, referred to Franklin's insight that lightning is a form of electricity as "the greatest, perhaps, that has been made in the whole compass of philosophy since the time of Newton" (Priestley 1767, 179).

As mentioned previously, Franklin's experiments with Leyden jars were extended to the manner of discharge with respect to differently shaped bodies. He found that it was easier to discharge a charged body by means of a pointed object than a blunt object. What's more, the discharge by means of a pointed object was accomplished with ease, showing no signs of the typical sparks and crackles. Franklin used this new knowledge to propose the erection of lightning rods on tall buildings, churches, and the masts of ships to draw the electricity from the heavens so that it could be discharged harmlessly into the Earth or ocean.

Franklin's proposal was dismissed by some theologians who insisted that thunder and lightning were tokens of divine wrath, and that it was therefore impious to interfere with their power of destruction. Defenders of Franklin's proposal countered that "it is as much our duty to secure ourselves against the effects of lightning as against those of rain, snow, and wind, by the means God has put into our hands" (cited in Benjamin 1895, 592). The lightning rod worked very well—so well that Franklin became celebrated among the scientists of Europe, the first scientist of the New World to enjoy this distinction.

The city of Brescia, Italy, was devastated in 1769 when the Church of San Nazaro, near Venice, was struck by lightning. The resulting fire ignited 200,000 pounds (90,000 kilograms) of gunpowder being stored there, causing a massive explosion which destroyed one-sixth of the city and killed 3,000 people. The disaster impelled the Roman Catholic Church to abandon its religious objection to using lightning rods to protect property.

By this time, the leading phenomena of the science of electrostatics were well known. There were two kinds of electricities, positive and negative, or perhaps only one, but it was such that it could be added or drawn from a neutral body. It was also known that electricity was conserved—that the sum of positive and negative charges was constant. There were insulators in which electricity could not move and conductors in which it moved freely. Equal charges repelled each other, whereas opposites attracted. With these fundamental facts understood, the time was rife for establishing a quantitative law for attraction and repulsion.

A QUANTITATIVE LAW FOR ATTRACTION AND REPULSION

The art of performing electrical demonstrations developed rapidly during the eighteenth century, but there was very little progress in actual measurement of the forces acting between charged bodies. Though Franklin and others had helped to clarify the notion of electric charge (quantity of electricity), all experiments and theoretical explanations to this point were purely qualitative.

One exception was the publication of *A Treatise of Artificial Magnets* in 1750 by John Michell (1724–1793), a Fellow of Queen's College, Cambridge. This treatise contains the first attempt to establish the inverse square law of force between magnetic poles. With the benefit of hindsight, this seems an obvious route for investigators to have taken in light of the importance of the inverse square proportion in Newtonian science, but scientists did not explore this possibility. Unfortunately, Michell's assertion that magnetic action conformed to the inverse square proportion was not persuasive because it was not backed by experiment.

The transformation of studies of electricity and magnetism into quantitative sciences was largely due to the work of Charles Augustin de Coulomb (1736–1806), a military engineer who was interested in a cluster of problems concerning the mechanics of machinery (strength of materials, rigidity, friction, and the like). He was drawn to a prize offered by the Académie Royale des Sciences in 1777 for the best design for a ship's compass, a competition that was uniquely suited to his knowledge of the properties of materials, especially the property of torsional rigidity. His memoir *Theorie des Machines Simples* (1779: *Theory of Simple Machines*)—a study of friction in machiney—won him first prize and membership into the Academy in 1781. Experiment, he asserts in this paper, reveals that the deflection of a magnetic needle cannot be caused by the flow of a liquid, but is occasioned by attractive and repulsive forces of the kind treated by Newtonian physics.

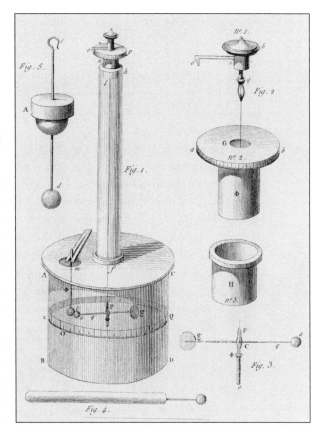

Figure 5.2: Coulomb's torsion balance. From Charles Coulomb, "Construction et usage d'une balance electrique" (1785).

In this paper, Coulomb illustrated a method for precisely determining the periods of oscillation of a magnetic needle, one that would allow the momentum of the magnetic force to be ascertained. At the same time, he introduced a small error that he was keen to eliminate. It was this small error that prompted a return to his studies of torsion in 1784 and the construction of a delicate torsion balance (fig. 5.2), which could detect a force equivalent to about 1/100,000 of a gram. The glass cylinder, marked *A B C D*, was covered by a round disk having one hole in its center and another at the side. Above the center hole projected a slender cylinder upon the top of which rested a graduated torsion micrometer, *op*, and suspended centrally from this was a fine silver wire, *qp*. This wire could be zeroed by turning the top torsion head. Attached at the lower end of the suspended wire was a carefully balanced straw covered with wax. The straw had a pith ball, *a*, attached to one end, and a small paper disk, *g*, set radially at the other. The disk counterbalanced the ball and dampened reed oscillations.

Suspended in this manner, the reed was free to rotate in a large glass cylinder that was marked in degrees around its circumference. Another charged ball was now lowered into the second hole, m, so as to be near the first ball. The operation of the apparatus depended on the principle that by similarly charging the two pith balls a repulsion force caused a twist of the wire that gave rise to a restoring couple measurable by the angle of displacement twist.

The device eliminated the small error that had crept into his award-winning paper of 1773. Although the device was difficult to use (his wire kept twisting around itself), Coulomb was able to measure precisely the small force that existed between the charges and determine how the force varies with variations in the amount of charge and the distance between them. The particles of electricity, he noted, behaved exactly like particles of ordinary matter. This result, which is known as Coulomb's law, states that the electrical force between two charged objects is directly proportional to the product of the quantity of charge on the objects and inversely proportional to the square of the separation between the two objects. In equation form, Coulomb's law can be stated as

$$F = k \frac{Q_a Q_b}{r^2}$$

where F is the magnitude of the force exerted, $Q1$ represents the quantity of charge on object 1 (in Coulombs), $Q2$ represents the quantity of charge on object 2 (in Coulombs), and r is the distance of separation between the two objects (in meters). The symbol k is a proportionality constant known as the Coulomb's law constant. The value of this constant is dependent upon the medium in which the charged bodies are immersed, and is often written as $\frac{1}{4\pi e_0}$.

Coulomb's law is analogous to the gravitational force in a number of respects. First, Newton's formulation is proportional to the mass of the objects, whereas Coulomb's is proportional to the charge. Furthermore, in conformity with Newton's third law of motion, the electrical force of one body exerted on the second body is equal to the force exerted by the second body on the first. Second, both the forces are inversely proportional to the square of the distances.

> ### The Electrical Researches of Cavendish
>
> Unknown to Coulomb, this law had already been recognized (but not published) by Henry Cavendish (1731–1810), arguably the greatest experimental scientist of the eighteenth century. Born into one of the greatest fortunes in England, Cavendish was an eccentric who lived alone, shunning society and especially women. Although he was elected a Fellow of the Royal Society in 1760, he kept his contacts with other scientists to a bare minimum. His place in the history of science is secured by virtue of his work in chemistry that, unlike his work in electricity, was published. Most of his results in electricity came to light a century later through the efforts of James Clerk Maxwell, who undertook the publication of Cavendish's unpublished electrical work during his term as Cavendish Professor of Experimental Philosophy.
>
> In the unpublished work, Maxwell discovered experimental proof of the inverse square law. Maxwell also discovered that Cavendish had anticipated Michael Faraday in demonstrating that the capacity of a condenser depends on the substance that is inserted between its plates. What's more, Cavendish used the concept of potential (called "degree of electrification"), which was a staple in the mathematical literature but not previously used in association with electrical experiments, in elaborating the hypothesis that all points on the surface of a good conductor are at the same potential with respect to a common reference, the Earth. This concept, first made plain by Cavendish, would play a vital role in the further refinement of electrical theory. Finally, in a series of experiments on various conductors, Maxwell found that Cavendish had anticipated the work of Georg Simon Ohm by showing that the potential across conductors was directly proportional to the current through them. Cavendish's finding was all the more remarkable given that he did not possess a way of measuring current. As a last resort, he used his own body as a meter, estimating the intensity of a current by gripping the ends of the electrodes with his hands and noting whether the shock terminated in his fingers, wrists, or elbows.

These similarities are offset by some differences. The Coulomb force can be attractive or repulsive, whereas the gravitational force is attractive only. Second, the magnitude of the Coulomb force is dependent on the medium separating the charges, while the gravitational force is independent of the medium. Finally, the Coulomb force constant k is numerically much larger than the gravitational constant, which means that for objects with charge that is of the order of a unit charge (C) and mass of the order of a unit mass (kg), that the electrostatic forces will be so much larger than the gravitational forces that the latter force can be ignored.

This result and others that constitute the foundation of the modern science of electricity and magnetism were published between 1688 and 1691 in a series of six memoirs by the Académie des Sciences. The first three of these papers were read at the Academy in 1785. Coulomb's first memoir dealt exclusively with repulsive forces. In the second memoir, Coulomb extended his observations to the interactions between spheres with positive and negative charges, as well as to magnetized bodies. Because the balls of his torsion balance kept sticking together, he modified his apparatus. His method was to suspend a magnetic needle from the torsion balance a fixed distance from another needle fixed upon a stand. The torsion arm was then deflected, and the resulting oscillations were timed. This procedure was repeated for varying distances between the oscillating and the fixed needles. In this way, Coulomb demonstrated that the forces responsible for motion were proportional to the inverse square of the period, and the period varied directly as the distances between the magnetic bodies.

Although two magnetic masses acted on one another in the same way as particles of electricity—that is to say, electrical and magnetic fluids acted according to formally identical laws—Coulomb did not conclude that the electric and magnetic fluids were identical. Electrical fluids move between the particles of ponderable matter, much as water and other fluids penetrate the pores of sponges. The two kinds of magnetic fluids, in contrast, do not circulate between the particles of ponderable matter but are confined within the molecules of matter. To explain magnetization, Coulomb theorized that half of a molecule of some substance, say iron, will contain a certain amount of the one kind of magnetic fluid, and the other half of the molecule will contain the same amount of the other kind of magnetic fluid. Magnetization occurs when these dipolar molecules are aligned so that their similar poles point in the same direction. Coulomb suspected that all molecules of ponderable matter contained the magnetic fluids, but his attempts to detect this universal magnetism were not successful. This theory became the standard model for magnetic materials throughout most of the nineteenth century.

In the third memoir of 1785, Coulomb examined the phenomenon by which electrified bodies gradually lose their charge. He showed that the leakage of electricity is proportional to the charge. He believed that the explanation for this phenomenon was direct contact on a molecular level between the charged body and its environment—charge was shared with molecules of air that surrounded the charged body.

Of the three remaining papers, the sixth memoir of 1791 is the most pertinent to our discussion to this point. In it, Coulomb discusses the various theories of the electrical fluid, with special attention to the debate between the one- and two-fluid theories of electricity. Although Coulomb's law defined the electric charge and he was the first to measure this charge, he was unable to define its nature. Expressing some misgivings, Coulomb sided with proponents of the latter theory. The two-fluid theory, he contended, is the most congenial theoretical framework for his many experiments and calculations. On no account, Coulomb insisted, should this theory be taken to explain the true causes of electricity.

Coulomb's position won the day everywhere but in England, America, and in Italy (where it was opposed by Volta). As we shall see in the next chapter, the discovery of current electricity through the invention of the voltaic pile in 1800 served only to reinforce the fluid concept. For the proponents of the two-fluid theory, a *current* rapidly came to mean the simultaneous flow of the two electrical fluids in opposite directions through a wire.

THE COMPLETION OF ELECTROSTATICS

By the end of the eighteenth century, the basic principles of electrostatics were nearing completion. The concept of electric charge and the principle of the conservation of electricity had been stated by Franklin. Coulomb had established the inverse square law governing the interaction of electric charges.

With these achievements in place, the great mathematical physicists of the day developed the theory of electrostatics in a manner that has stood up to the present day. The first step was the publication in 1759 of a sophisticated mathematical treatment of magnetism, and electrical phenomena that were analogous to magnetism, based on Franklin's one-fluid model by the German Franz Aepinus. The distinguishing feature of Aepinus's *Tentamen Theoriae Electricitatis et Magnetismi* (*An Attempt at a Theory of Electricity and Magnetism*) was its employment of algebraic analysis.

In 1777, the mathematician Louis Lagrange (1736–1813) condensed and simplified the theory of attractions by showing that the components of the attractive force at any point can be simply expressed as derivates of the function obtained by adding together the masses of all the particles of an attracting system, each divided by its distance from the point. Five years later, he showed that this function $V(x, y, z)$ satisfies a partial differential equation in space free from attracting matter and is equivalent, in three dimensions, to the inverse square law of gravitational or electrical attraction. In 1812, Siméon Denis Poisson (1781–1840), a mathematical physicist, put the finishing touches on the mathematical apparatus of Coulomb's theory: He borrowed the potential (V) from Lagrange and wrote the corresponding differential equation ($\Delta V + 4\pi p = 0$, where p is the charge density), solved this equation, and improved its agreement with Coulomb's experimental results.

Coulomb's electrostatics harmonized beautifully with the scheme that dominated French physics, associated most readily with the towering figure of Pierre Simon Laplace (1749–1827), which attempted to reduce every physical phenomenon to central forces acting between particles of ponderable and imponderable fluids, in analogy with gravitation theory. The electric fluids were controversial elsewhere: In Britain, the single-fluid theory dominated.

Just as the science of electricity was maturing into what appeared to be a complete body of knowledge, new experimental discoveries would disclose that scientists were only cognizant of a small portion of the phenomena. The new departure, as we shall see in the coming chapter, would come from a completely unexpected field of inquiry—human anatomy.

6

THE SCIENCE OF GALVANISM

INTRODUCTION

Electric charge can move from one point to another—a phenomenon that can also be described as the flowing of an electric current.[1] However, prior to 1800 scientists were only able to produce nearly instantaneous flows of this sort. For example, electric charge could be transferred from a Leyden jar to a human body, or to some other receptacle, but this transfer was accomplished in one quick spark or jolt. To produce a continuous transfer of charge from one point to another, scientists had to create a device that would produce a new supply of charge at point A as fast as it is moved to point B, and in turn, consume it at point B as fast as it is transferred there.

The invention of just such a device—the first continuous source of electric current—was announced in 1800 by Alessandro Volta (1745–1827), professor of physics at Pavia. Volta's electric battery ushered in an entirely new type of science called "current electricity," or the science of electrodynamics. Where electrostatics studied electric charges, the way they separate and combine, the attraction and repulsion between charges, and the invisible electrical fields they create, current electricity is the study of the flowing or wiggling of charges, of the magnetic attraction and repulsion they create, and of the way they can move energy around.

GALVANI'S ANIMAL ELECTRICITY

Volta was stimulated by the research of Luigi Galvani (1737–1798), an anatomist and physician at the world's oldest continuing university, the University of Bologna. From very early times, it had been known that some aquatic animals have the ability to inflict a shock. For example, the electric eel that is found in the Basin of the Amazon River can generate electrical shocks of up to 600 volts, which it generates through stacked electroplaques, in a manner similar to a battery. Following the invention of the Leyden jar, scientists became interested in the relationship, if any, between the physiological effect of its discharge and that of the shock given off by these aquatic animals.

In a series of experiments that commenced around 1780 and consumed the better part of a decade, Galvani found that dissected frogs legs twitched as though in convulsion when they were placed in contact with a spark from an electric machine. To his astonishment, he also found that a metal scalpel caused the frog's legs to twitch if the machine was turned on, even if the spark didn't actually make contact. Galvani varied his experimental design as much as possible. He substituted a Leyden jar for the rotating electrical machine. He tried contacting the nerves of the frog's legs with an assortment of materials. He went so far as to use different sorts of animals, some warm-blooded, others cold-blooded.

If an electrical spark caused this muscle twitching, Galvani reasoned that he could confirm Franklin's contention that lightning was indeed a form of electricity. To test Franklin's hypothesis, Galvani hung the legs of frogs by their nerves from brass hooks against an iron latticework (fig. 6.1). The lower tip of the suspended member was connected to a grounded wire. He found that the legs twitched whether a thunderstorm was in the vicinity or not. He noticed that when a copper or brass hook had been pressed into the frog's marrow and hung from an iron trellis, twitching would be seen even when the weather was

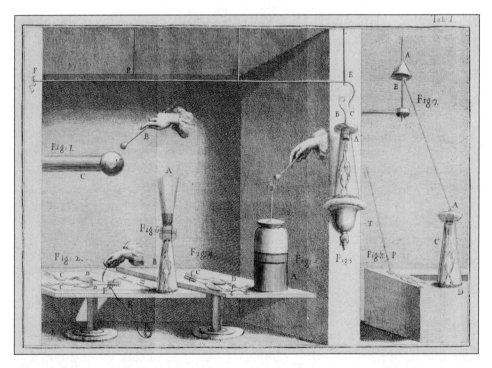

Figure 6.1: Galvini's experiment on frogs. From Galvani, *De viribus electricitatis in motu musculari* (1791).

pleasant. The twitching occurred, he discovered, whenever the muscles came into contact with two different metals at the same time. Atmospheric conditions were clearly not the principal cause of the muscular contractions. To confirm this suspicion, he brought the specimen indoors. Using an iron plate instead of the trellis, he again observed the same twitching.

Galvani still wasn't sure how to explain the result of his experiments. He considered two possibilities. The first was that the metals caused the twitching. If this hypothesis was sound, the legs simply indicated an electric charge generated from outside: They would show the stimulation just as an electroscope would show an electrical charge by the divergence of the straws or metallic ribbons. The second possibility was that the muscles retained some sort of innate electricity even after the animal had died. Galvani opted for the second interpretation. Conceiving a living creature as a fleshy kind of Leyden jar, he decided that the motion was caused by the transport of a peculiar *fluid* from the nerves to the muscles. In Galvani's own words: "It would perhaps not be an inept hypothesis and conjecture, nor altogether deviating from the truth, which should compare a muscle fiber to a small Leyden jar, or other similar electric body, charged with two opposite kinds of electricity; but should liken the nerve to the conductor, and therefore compare the whole muscle with the assemblage of Leyden jars" (Galvani 1791). To this peculiar fluid, the names *galvanism* and *animal electricity* were generally applied.

In Galvani's theory, the nerves and muscles of the frogs in his experiment constituted the inner and outer charged surfaces of the jar. When the outer surface of a muscle received an electrical charge (like the outer surface of the Leyden jar), the nerve and inner muscular surface became oppositely charged and muscular contraction followed.

THE INVENTION OF THE ELECTRIC BATTERY

Galvani's paper "De viribus electricitatis in motu musculari commentarius" (Commentary on the Effects of Electricity on Muscular Motion), which was published in the *Proceedings of the Bologna Academy of Sciences* in 1791, generated a great deal of discussion. Galvani also printed his paper as a pamphlet and circulated a dozen or so copies among his fellow scientists. Granted that his studies seemed to call for a bold new conception of life itself in terms of an electric fluid that animated nature, the hubbub surrounding his research was not surprising. Some scientists endorsed Galvani's own view that galvanism was the same as ordinary electric fluid. Others held that galvanism was a fluid distinct from ordinary electricity. And a third position refused to attribute galvanic phenomena to a fluid found in the nervous system.

One recipient of Galvani's paper was Alessandro Volta, who had established his reputation through the invention of the electrophorous (chapter 5). On the basis of his work with static electricity, Volta was appointed professor of physics at the University of Pavia, a post that he retained until his retirement. It was at Pavia that he made his most important findings.

Volta received Galvani's paper when he was 45 years old. His initial reaction was to dismiss these experiments on animal electricity as "miraculous" and "unbelievable"— a reaction that was perhaps nurtured by his professional distain for physicians. At the insistence of his colleagues in pathology and anatomy at his own university, he carefully repeated Galvani's experiments. At first, he accepted Galvani's conception of the frog as a fleshy Leyden jar, but around 1793 or so he began to suspect that the frog is primarily a detector of electricity and that the cause was to be found outside the animal.

Volta's repetitions of Galvani's experiments increasingly focused on Galvani's model of a Leyden jar. Volta concentrated the metallic probes on the nerves only, taking the muscles out of the picture. By taking the muscles out of the picture, the experimental arrangement no longer functioned analogously to the Leyden jar. Volta found that the redesigned apparatus produced the same results as Galvani's experimental arrangement.

Volta noted that electricity was produced only if two different metals were used; if the circuit consists of one and the same metal, the frog muscle did not contract. He also noticed that some combinations of metals produced more twitching than others. He tried placing a piece of tinfoil and a silver coin in his mouth, one on top of the tongue, the other touching his tongue's lower surface. When the tinfoil was pressed to the coin, he found that this apparatus produced a sour taste in his mouth, which he interpreted as indicating the presence of an electrical discharge. The taste persisted as long as the tin and silver were in contact with one another, showing that the transfer of electricity from one place to another was continuous. The German professor of mathematics Johann Georg Sulzer (1720–1779) had anticipated Volta's work by discovering in 1752 that if two pieces of metal, one of silver and the other of lead, were joined so that their edges touched, and are then placed on the tongue, a taste similar to "vitriol of iron" (a sulfate of iron) was perceived. Unlike Volta, Sulzer did not connect this experiment with the production of continuous current electricity.

Volta then substituted a silver spoon for the coin and a copper wire for the tinfoil, with the same result. When he touched the end of a strip of one metal to his forehead and the end of a different metal to his palate, he claimed that a bright flash seemed to appear in his mind when the free ends of the strips touched. The metals, he concluded, were not just conductors but were actually responsible for the production of electricity. The frogs' legs had exhibited not animal electricity, as Galvani had supposed, but metallic electricity.

Volta realized that if the nerves and muscles of animals simply served to indicate the presence of electrical energy, he could use a condensing electroscope to give a more accurate measurement of the electricity. He prepared a number of plates of different metals, each furnished with an insulating handle so that when one was placed on the other, an electrical charge would be generated. When the charge was measured by the electrometer, the device indicated whether it was a positive or negative charge. He found that when a zinc disk was brought in contact with a copper disk, the zinc disk became positively charged and the copper became negatively charged. By carefully gauging the effects on the diverging gold leaves of his electrometer, Volta arrived at the following table of metals in the order of their relative charges, each with respect to the other:

POSITIVE	NEGATIVE
Zinc	Copper
Lead	Silver
Tin	Gold
Iron	Graphite

Continuing his experiments, Volta realized that electrical forces were generated not only when two dissimilar metals touched, but also when a metal touched certain kinds of fluids. The electricity might be conducted from one metal to the other by the fluid in his saliva. When he placed disks of silver, tin, or zinc on moist cloth, clay, or wood, and then separated them and brought them to the electrometer, a negative electrification resulted. He showed further that improved results were obtained when a circuit was formed by two different metals separated by a moist element, and that their effectiveness was added together when such combinations of metals and moist elements were stacked in a repeating pattern.

In this way, Volta succeeded in building his famous column, or pile, of electric generating elements. He experimented with many solutions, finally settling on brine, which is a strong solution of saltwater. He found that if he constructed piles, or couples, of dissimilar metal disks, sandwiching pieces of brine-soaked cardboard in between them, he had an effective set of battery cells, each couple forming one cell (fig. 6.2). (The term *electric battery* was first used by Benjamin Franklin, who applied it to a row of Leyden jars joined together to increase the electromotive force of the collective electrical discharges.) The resulting stack became known, after its inventor, as the *voltaic cell*. With this new device, there was no instantaneous flash, as seen in a frictional electric discharge through an air gap.

In contrast to the older frictional electricity, with high potential or driving force (voltage) and little current (amperage), the new galvanic electricity

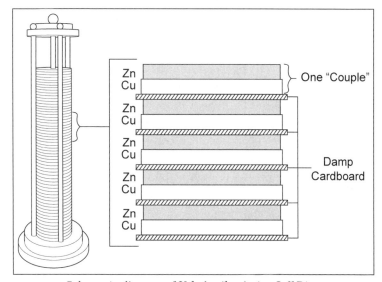

Figure 6.2: Schematic diagram of Volta's pile. Artist: Jeff Dixon.

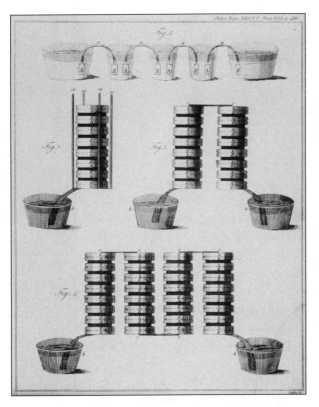

Figure 6.3: Illustration of two forms of Volta's pile. From Volta, "On the Electricity Excited by the Mere Contact of Conducting Substances of Different Kinds" (1800). Courtesy of Smithsonian Institution Libraries.

possessed low voltage but enormous current. Volta used the somewhat dated term *electrical tension* for the more familiar *electrical potential*. In modern electrochemistry, the term *electrical potential* is ordinarily restricted to the difference between the two electrodes of an electrochemical cell when no current is flowing through it. The term *voltage* generally refers to the same with current flowing.

Volta devised two forms of his voltaic cell. The *couronne de tasses* form is depicted in the upper part of figure 6.3. It featured a ring of cups filled with brine, connected by alternate strips of zinc and silver joined by metallic jumpers. The columnar apparatus appears in the middle and lower parts of the plate. It featured a pile of silver and zinc disks, each pair of which was separated from the adjoining pairs by paper or cloth separators soaked in brine. The former represented an improvement on the latter because the current tended to weaken as the brine in the separators dried out.

Volta was not clear on why his new device was a source of continuous current. We now know that this phenomenon occurs because atoms of all matter include, as a part of their internal structure, negatively charged electrons and positively charged protons. Indeed, the electric charge produced by a continuously operating Voltaic pile is not being created by the arrangement of metal disks but is present in matter all the time. For the pile to work, all that is required is that the arrangement of disks separates the already-existing negative and positive changes—a separation that is most simply achieved when the pile involves only two different metals. If we take two metals from Volta's list, say zinc and copper, each metal contains electrons bound by forces of greater or lesser extent to the atoms of the metal. The forces binding the electrons to zinc atoms are somewhat weaker than those binding electrons to copper. At the boundary, electrons tend to slip across from zinc to copper. The copper, with its stronger grip, tends to wrestle the electrons from the zinc.

Volta, however, fell back on the theory of an electrical fluid to explain the working of the pile. Volta (1800, 403) hypothesized that statical electric fluid was somehow liberated and forced into motion at each zinc-copper interface. This electric fluid was driven from the first couple, freely traversing the cardboard layer, receiving an additional impulsion at the copper-zinc interface of the next higher couple, and so on.

Volta demonstrated the action of his new *"organe électrique artificiel"* in Paris in November 1801, before an assembly of scientists. The demonstration was attended by Napoleon Bonaparte, himself a member of the Institut de France, who personally assisted with the experiments. Napoleon drew sparks from the pile, melted a steel wire, discharged an electric pistol, and decomposed water into its elements. For his achievement, Volta was awarded a special gold medal and given a pension.

THE NEW SCIENCE OF ELECTROCHEMISTRY

Volta dispatched a report of his experiments to Sir Joseph Banks, president of the Royal Society in England. His report was written in French, but it bore the English title "On the Electricity Excited by the Mere Contact of Conducting Substances of Different Kinds." It was published in volume 90 of the *Philosophical Transactions of the Royal Society* in 1800.

Scientists on both sides of the Atlantic immediately set out to replicate Volta's experiments, for it was clear that a source of constant-current electricity not only opened up entirely new avenues of research, but had the potential to transform other established disciplines in ways that had previously been unimaginable. The discipline that reaped the greatest immediate benefit from Volta's battery was chemistry, which now possessed a powerful new tool for tracking down new elements and for coming to grips with the elusive nature of chemical bonding. Antoine Lavoisier (1743-1794), the architect of a new chemical philosophy that had revolutionized the practices of chemists, had defined an element as a substance "so far undecomposed." The system of chemistry, introduced by Lavoisier in his monumental *Traité Élémentaire de Chimie* (1789: *Elementary Treatise of Chemistry*), placed such substances as lime and magnesia among the elements, but its author had prognosticated that the compound nature of these earths would one day be revealed. Lavoisier was so confident that potash and soda were compounds, however, that these substances were omitted from the table of elements prepared for the *Traité* (Lavoisier 1789, 175-178).

Even before the second part of Volta's historic announcement reached the Royal Society, the chemist William Nicholson (1853–1815) and the surgeon Anthony Carlisle (1768–1840) had constructed the first pile made in England and placed in the service of chemistry. The first notable result, reported in Nicholson's *Journal* for July 1800, was that water was decomposed by their pile, which consisted of 17 large silver half-crown pieces alternated with equal disks of copper and cloth soaked in a weak solution of ordinary salt. Gas was observed to bubble at the junction of the conducting wire and the pile, where a drop of water had been added to improve the conductivity of the junction.

Nicholson and Carlisle later improved this crude apparatus by using a glass tube filled with water and with brass wires connected to the pile terminals, entering each end of the tube through a cork stopper. Hydrogen formed at the negative terminal and oxygen at the positive one. Nicholson and Carlisle were not the first to decompose water with electricity, but they were the first to study the phenomenon systematically and to prove that the separated gases were hydrogen and oxygen. Their work made it clear that electricity was related to chemical affinity, and that the sciences of electricity and chemistry must have intimate and hitherto unforeseen connections. As for Volta, he was not at all interested in the chemical effects of the pile, but dedicated his efforts toward refuting Galvani's theory of animal electricity, which still enjoyed a great deal of support.

These experiments were repeated by the English chemist William Cruickshank (1745–1802), who had discovered the metal strontium in 1787. He worked with an improved version of Volta's battery. He arranged square sheets of copper, which he soldered at their edges to similar sized sheets of zinc. These couples were placed into a long rectangular wooden box lined with resinous cement. The couples were set into grooves and an electrolyte of brine or dilute acid was added to fill the box. This arrangement, known as the *galvanic trough,* avoided the drying out of the spacers in Volta's battery and also provided a much more powerful source of electric current. This design was the basis for a powerful battery of 600 pairs presented by Napoleon Bonaparte to the Ecole Polytechnique in 1808.

With this arrangement, Cruickshank showed that hydrogen evolved from the silver or copper section of the pile, and oxygen from the zinc. He also decomposed several compounds and electroplated metals. He decomposed the chlorides of magnesia, soda, and ammonia, and he was also able to precipitate pure copper and silver from their salt solutions—a process that led to the beginnings of the great metal refineries. A powerful new tool was replacing the electrostatic generators of the late 1700s. Sparks, halos, and pinwheels gave way to the more challenging phenomena that issued from the ends of an electric battery assembly.

THE DISCOVERY OF NEW ELEMENTS

The most ambitious experimenter with the new technology was Humphry Davy (1778–1829), who is now best known for his discovery of two elements, potassium and sodium, as well as for his invention of a safety lamp for miners. In 1798, Davy accepted a position at Beddoes Pneumatic Institution in Bristol, working on the therapeutic uses of gases. He discovered how to prepare nitrous oxide (laughing gas), and prepared a well-reasoned paper on it that established his reputation as a chemist. His laboratory was frequented by several leading social figures, including the poets Coleridge and Wordsworth, who enjoyed the intoxicating effects of his discovery.

Davy's paper caught the attention of Count Rumford (Benjamin Thompson, 1753–1814), who had recently founded the Royal Institution in London (see the Postscript at the end of this chapter for more about the Royal Institution). Rumford hired Davy as the first director of the laboratory at the Royal Institution, and under his leadership it became a center for chemical research and popular expositions of science. To raise money, the institution developed a highly popular lecture series, and Davy was one of the most sought-after lecturers of his day. More than 1,000 people attended one of Davy's lectures in 1810. To pursue the ideal of popularization, the institution focused on agricultural science, tanning, and mineralogy. A book on agricultural chemistry and several excellent papers on these subjects by Davy added to the stature of the Royal Institution and to his own reputation as a scientist.

Davy's first lecture at the Royal Institution on April 25, 1801, presented a history of galvanism. In collaboration with William Hasledine Pepys (1775–1856), in February 1802 Davy constructed the strongest voltaic pile of its day. It consisted of 60 pairs of zinc and copper plates, each 6 inches square, held in 2 large troughs filled with 32 pounds of water containing 2 pounds of azotic (nitric) acid. With this device, Davy melted iron wires up to one-tenth of an inch in diameter, culminating in an exceedingly bright light, and platinum wires one-thirty-second of an inch in diameter that turned to white heat and melted in globules at the point of contact. Gold, silver, and lead sheets were made to glow and ignite.

As the title of Volta's letter to Banks testifies, Volta contended that the voltaic cell generated electricity by the mere contact of dissimilar metals. Indeed, he had reached this conclusion by 1796. In his view, the contact between silver and tin, for example, gave rise to a force that caused the first metal to give electrical fluid and the second to receive it. If the circuit contained moist conductors (the electrolyte), this force would produce a current or continuous flow of the fluid, which would travel from the silver to the tin, and from there via the moist conductor back to the silver, and then back to the tin, and so on. If the circuit was not complete, the result would be an accumulation of electrical fluid in the tin. Although the electrolyte was required for the production of current, Volta stressed that metal-to-metal contact was the seat of the electricity. Volta was not clear on what caused the charge separation, but he did offer an expression to describe it—*electromotive force (EMF)*— which he defined in 1801 as a measure of the disturbance of the equilibrium of electricity between two metals, equal to the tension in an open circuit.

Volta's contact theory was opposed by the many chemists who were persuaded that chemical processes were the very cause of the pile's actions. Davy persuaded himself of the inadequacy of Volta's contact theory, convinced that the source of voltaic electricity was a chemical action. The seat of the electromotive force might still be at the points of metallic contact, but it was clear to Davy that a transformation of energy took place in the voltaic cell. Davy conjectured that the reverse might also be true—that the application of electricity to compounds and mixtures might produce chemical reactions; in short, many substances that had resisted chemical attempts to split them apart might now be separated by electricity. In the space of five weeks in 1806, Davy performed 108 experiments in electrolysis (the use of electricity to produce chemical changes). These experiments produced a key theoretical insight—namely, chemical affinity is electrical. This insight was later developed into the dualistic system, whereby chemical elements were regarded as electrically positive or negative, which combined to form neutral products that could be decomposed by an electric current. However, it was not part of Davy's makeup to elevate his insights into formal systems, and so this task fell to his successors.

For years, scientists had attempted to break the bonds between parts of compounds to isolate their constituent elements. Lime, magnesia, potash, and others that seemed to be oxides of metals had been the target of these efforts. No one had succeeded in breaking away the tightly held oxygen. In 1807, Davy constructed an especially powerful battery in an attempt to electrolyze two substances that were long suspected of harboring metals as part of their chemical structure—caustic soda (sodium hydroxide) and caustic potash (potassium hydroxide). His battery consisted of a trough lined with pitch and divided into a series of cells by more than 250 metal plates. The dividers were made by joining together plates of zinc and copper face to face, so that each division was in fact a double plate—zinc on one side and copper on the other. He poured acid into the gaps between the plates, producing a series of cells that, when connected together, made the most powerful battery yet constructed.

On October 6, 1807, Davy passed the current from this battery through a lump of dampened potash (made by soaking the ashes of burnt plants in pots of water). At both points of contact, the potash began to fuse, giving off a gas on the surface that was attached to the positive pole. At the other contact point, no gas was given off but small globules with a high metallic luster began to form. These globules looked a lot like droplets of mercury, and some of them burned brightly and exploded. Davy knew that he had discovered a new element, which he called potassium (after "potash"). A week later he utilized this same method to isolate sodium from caustic soda. Davy's discovery of sodium and potassium shook the very foundations of the French school of chemistry. If the oxides of sodium and potassium formed the strongest alkalies, then Lavoisier's contention that oxygen was the principle of acidity was clearly mistaken.

In Stockholm, the chemist Jöns Jacob Berzelius (1779–1848) and his colleagues were pursuing similar experiments. Berzelius suggested that chemical affinity arises from the play of electric forces, which in turn spring from electric charges within the atoms of matter. Each atom, he suggested, is composed of two poles, which are the seat of opposite electrifications and whose electrostatic field is the cause of chemical affinity. The picture that emerged of chemical combination on this theory was this: Two atoms that are about to unite dispose themselves so that the positive pole of one touches the negative pole of the other. The electricities of these two poles then discharge each other, giving rise to the heat and light that are observed to accompany the act of combination. When the heat and light disperse, the compound molecule is left behind, with its two remaining poles. This compound cannot be disassociated into its constituent atoms again until some means is found of restoring to the vanished poles their charge. The galvanic pile provides just such a means: The opposite electricities of the current invade the molecules of the electrolyte and restore the atoms to their original state of polarization.

Berzelius had found that he obtained an "amalgam," or alloy, of some other metal with mercury when he ran a current through a mercury compound added to lime or baryta (barium oxide or hydroxide). This result furnished Davy with another clue. He constructed a new battery with 2,000 pairs of plates of zinc and copper. It produced the most brilliant arc yet. Within a few months, by applying a strong current to the amalgams described by Berzelius, Davy isolated magnesium (from magnesia), calcium (named after "calx," the ashy powder sometimes caused by heating a chemical substance), strontium (from a mineral named for a Scottish town called Strontian), and barium (from baryta).

Davy also tested a green gas that was given off when muriatic acid was decomposed (called oxymuriatic acid), recognizing it in 1810 as an element he called chlorine (for its greenish color). He ignited cotton, sulfur, resin, and ether. Iron, quartz, sapphire, and platinum were melted; diamond was evaporated; and liquids, such as water and oil, were boiled by the electric current. The intense heat developed by the battery was used in 1815 to melt a short section of iron wire and diamond dust together, thereby carburizing the iron and forming steel.

The electrochemical experiments of Davy and Berzelius, which led to the discovery of new earth metals, to the plating of metals by electricity, and to the brilliance of arc light, were not the only practical results of Volta's great contribution. Another result that proved to be particularly important was a mechanism for exploding gunpowder. The standard practice at the time was to ignite gunpowder by means of a fuse, but it was a short step to using a spark instead. In 1812, Baron Paul L. Schilling, attaché to the Russian Embassy in Munich, insulated a copper wire with a thin coat of India rubber and varnished and stretched this wire underground and under water so as to explode powder mines across the river Neva near St. Petersburg. This was also done across the Seine during the occupation of Paris by the allied armies following the collapse of Napoleon. The latter demonstration was an astonishing sight to those who first witnessed it but did not understand the technique used.

THE END OF AN ERA

At the close of the eighteenth century, the science of electricity and magnetism consisted of three separate fields—electrostatics, galvanism, and magnetism. The most refined of these fields was electrostatics, the dominant theory of which was due to Coulomb, Laplace, and Poisson. Following in the footsteps of Franklin, in the mid-1780s Coulomb established the inverse square law for the repulsion between like electric charges and the attraction between opposite ones. Later, in 1812, Poisson produced a mathematical treatment of electrostatics based on Laplacian mechanico-molecular principles that would become the standard treatment of the subject, both in France and Britain. Assessments of the physical nature of electricity were split to a significant degree along national lines. English scientists followed Franklin in treating electricity as a single imponderable fluid, whereas the French followed Coulomb in endorsing the two-fluid theory.

The theory of magnetism was less advanced than that of electricity, but the inverse square law had been established by Coulomb, and in 1824 Poisson would publish a mathematical treatment of magnetism to complement the one that he had already provided for electrostatics.

The third area, galvanism, presented some difficult issues. The operation of the pile was not well understood. Volta had argued that the electricity produced in the pile was essentially an electrostatic phenomenon, and his contact theory had come to enjoy guarded support. Although Volta's theory was supported by Laplace, Poisson, and Biot, it was opposed by the chemical theory advanced by Davy and a number of important British scientists. There was still no mathematical theory of galvanism.

POSTSCRIPT: FOUNDING OF THE ROYAL INSTITUTION

The Royal Institution was founded in 1799 during a period of increasing industrialization for Britain. The stated purpose of the institution was "for diffusing the Knowledge, and facilitating the general Introduction, of Useful Mechanical Inventions and Improvements; and for teaching, by Courses of Philosophical Lectures and Experiments, the application of Science to the common Purposes of Life." The new institution was housed at a building at 21 Albemarle Street, which was converted into a proper scientific institution with laboratory space, a lecture theatre, libraries, and offices.

During the first few years of its existence, the Royal Institution was chiefly concerned with improving agriculture by the use of chemistry. Restriction on trade with European neighbors created a demand for improved agricultural techniques and the dissemination of these improvements to farmers. These utilitarian concerns continued through the nineteenth and twentieth centuries, but it soon became apparent that the institution needed broader financial support if its work was to continue. The answer was the inauguration of a series of scientific lectures and courses of instruction by public subscription. Humphry Davy and Thomas Young (1773–1829) were hired as full-time research scientists and lecturers. (Although Young enjoyed an enormous reputation as a scientist through his *Outlines of Experiments and Enquiries Respecting Sound and Light* (1800), which established the wave theory of light, his lectures were a failure and he resigned his post after a two-year stint.) Davy's lectures, which dealt with a wide range of scientific issues at the cutting edge of scientific research, proved to be so popular that the original agricultural role of the institution disappeared, and its day-to-day activities became increasingly dominated by scientific research. Davy's appointment as professor of chemistry ensured that scientific research would be a crucial function of the institution.

Michael Faraday was appointed as Davy's assistant in 1813 and eventually was appointed Fullerian professor of chemistry in 1833. He inaugurated many of the traditions that are still alive at the Royal Institution. He founded the Friday Evening Discourses—a series of formal lectures given to members and their guests on a wide range of scientific topics. It was during his discourse in April 1897 that J. J. Thomson (1856–1940) would first announce the existence of the fundamental particle later called the electron. Faraday also founded in 1826 the immensely popular lectures for children called the Christmas Lecture Series, which are still presented to this day. The aim of these lectures was to explain complex ideas to young people in a manner that would capture the hearts and minds of the young.

NOTE

1. Physics texts often make reference to "flows of current." Electric current is a flowing substance, but the name of the substance is not "current." An electric current is a "flow of charge." Electricity can flow and electricity can stop, and a flow of electricity is called an electric current, but there is no such thing as current electricity. Electric current is actually a flowing motion of charged particles. The expression "electric current" means the same as "charge flow."

7

THE CURRENT AND THE NEEDLE

IMMANUEL KANT'S PROGRAM

Isaac Newton's monumental *Principia* advanced not only three laws of motion and his celebrated law of universal gravitation, but also a set of methodological rules that were designed to assist scientists in their description of natural phenomena. In the first edition of 1687, these rules are called *hypotheses;* by the third edition of 1726, they are called *regulae philosophandi,* or rules of reasoning. Newton's third rule states:

> In experimental philosophy we are to look upon propositions inferred by general induction from phenomena as accurately or very nearly true, notwithstanding any contrary hypotheses that may be imagined, till such times as other phenomena occur, by which they may either be made more accurate, or liable to exceptions. Newton 1934, 400.

This rule stipulates that those qualities (e.g., inertia, impenetrability, and mass) that are found universally in all sensible matter may be assumed to be the qualities of the ultimate particles of which sensible matter is composed. It struck many scientists as lacking a sound foundation. Although he regarded himself as a good Newtonian, the German philosopher Immanuel Kant (1724–1804) was a prominent critic of Newton. Kant's greatest philosophical work, *Critik der Reinen vernunft* (1781: *Critique of Pure Reason*), insisted in opposition to Newton that it was impossible to know whether or not atoms even exist. It followed, he stated, that talk of their qualities merely piled speculation on conjecture. What we can do is describe nature and nature's laws. We cannot, however, offer any explanation of subsensible entities like atoms. Kant's criticism proved to be very influential on Auguste Comte (1798–1857) and others who developed the positivistic view of science, which held that the aim of science is simply to describe the phenomena experienced, not to penetrate to things thought to lie behind the phenomena.

In his later *Metaphysiche Anfangsgründe der Naturwissenschaft* (1786: *The Metaphysical Foundations of Natural Science*), Kant advanced a philosophically interesting approach to the problem. He asked: What do we really mean by matter? Surely, it is not little atoms bouncing around in empty space, for the distance between this idea and that of solid matter is almost infinite. What we actually mean, Kant submitted, is the resistance of an object to our attempts to move through it. A solid, for example, is merely a clearly defined space that prevents our fingers from penetrating it. A table pushes back on our hands when they rest on its surface, and a surface may be defined as a zone of repulsive forces resisting penetration. The pushing back is the basic property by which we recognize that such matter exists. But there is more to the table than repulsion. The table, after all, does not swell to fill all space, as it would do if it were merely a repulsive force in space. Any attempt to pull the table apart testifies to the existence of a force that holds the table together. The table that we experience, then, is merely the result of attractive forces holding in the repulsive forces within clearly defined boundaries. In the last analysis, reality consists of the two forces of attraction and repulsion. There is no need to assume a solid (whatever that means) substratum for force. The forces themselves suffice.

Kant held that attraction and repulsion are not confined to material objects. These forces occur in thermal, optical, electrical, and magnetic phenomena as well. It did not take long for Kant's followers to suggest that all those physical forces were merely manifestations of the fundamental attractive and repulsive forces under different physical conditions. All observable forces, therefore, were really only the appearance of the attractive and repulsive forces in different guises, depending on the circumstances.

A necessary corollary followed: All physical manifestations of force ought to be convertible one into another. If light, heat, electricity, and magnetism were merely the attractive and repulsive forces disguised as optical, thermal, electrical, and magnetic forces, then it should be possible to transform light, heat, electricity, and magnetism one into another. The mode of transformation appeared to follow from the forces themselves. Attraction and repulsion are polar opposites. There must exist a tension between them, and transformations occurred when the tensions became too great to be contained by any specific manifestation as electricity or magnetism. Thus positive (+) and negative (-), when combined, do not lead to annihilation but to transformation. Implicit in this circle of transformations is the conservation of the total force.

Although Kant's ideas were not very influential in England and France during the last two decades of the eighteenth century, in Germany they created an intellectual revolution. One of Kant's most ardent disciples was F.W.J. Shelling (1775–1854). His most important contribution was his recognition of what was only implicit in Kant. If matter and material phenomena could be seen only as the results of attractive and repulsive forces, then it seemed obvious to Shelling that all phenomena should be reduced to the force of attraction and repulsion. Kant hinted at this implication, but it was Shelling who elaborated on this view during the last decade of the eighteenth century.

THE DISCOVERY OF ELECTROMAGNETISM

By the close of the eighteenth century, electricity and magnetism were widely regarded as disconnected phenomena. Despite the mathematical analogy of their fundamental laws of equilibrium, it was thought that their causes and effects were completely distinct from one another: Electrification required a violent action and implied violent effects such as sparks and thunder, whereas magnetism was a quiet force. The magnetizing effect of thunder, which had been known for some time, was dismissed as a secondary effect of mechanical origin.

Another disciple of Kant, Hans Christian Oersted (1777–1851), studied at Copenhagen University, where he completed a doctoral dissertation in 1800 on Kant's philosophy. Later, he became a professor of natural philosophy at Copenhagen, was a student of *Naturphilosophie,* and was firmly committed to doctrine of the unity of forces. The discovery of the chemical importance of electricity was for Oersted confirmation of this position; he believed that chemical affinity was the chemical manifestation of the electrical force. It represented a conversion of force, not a peculiar action of a species of matter. Such a conversion, Oersted came to believe, ought to be merely the first in a chain of similar conversions in which one force could be turned into another under varying circumstances.

As early as 1806, Oersted began to look for other conversions. In 1813, in a work on the force of chemical affinity entitled "*Recherches sur l'identité des forces chimiques et électriques*" (Research on the Identity of Chemical and Electrical Force), he suggested that electricity ought to be convertible into magnetism under the appropriate circumstances:

> One has always been tempted to compare the magnetic forces with the electrical forces. The great resemblance between electrical and magnetic attractions and repulsions and the similarity of their laws necessarily would bring about this comparison. It is true, that nothing has been found comparable with electricity by communication; but the phenomena observed had such a degree of analogy to those depending on electrical distribution that one could not find the slightest difference ... The form of galvanic activity is halfway between the magnetic form and the electrical form. There, forces are more latent than in electricity, and less than in magnetism.... But in such an important question, we would be satisfied

if the judgment were that the principle objection to the identity of forces which produce electricity and magnetism were only a difficulty, and not a thing contrary to it.... One could also add to these analogies that steel loses its magnetism by heat, which proves that steel becomes a better conductor through a rise in temperature, just as electrical bodies do. It is also found that magnetism exists in all bodies of nature, as proven by Bruckmann and Coulomb. By that, one feels that magnetic forces are as general as electrical forces. An attempt should be made to see if electricity, in its most latent stage, has any action on the magnet as such. (Oersted 1813, 253)

Oersted spent the next seven years engaged in various activities. He continued as a popular lecturer and accepted the post of secretary of the Royal Danish Society of Sciences in 1815. Although these activities left him little time for the laboratory, he nevertheless pioneered a number of important innovations. A year later, he improved the voltaic battery, replacing the wooden trough with a copper one, which produced stronger currents. He also devised a detonating fuse in which a short wire was caused to glow by an electric current.

During a lecture in 1820, he passed electricity through very fine platinum wires. The association between electrical forces (both electrostatic and voltaic) and magnetic forces was well known. In the late 1700s, scientists had magnetized iron by sending an electrostatic charge through it. It was also known that thunderstorms often affect the direction that a magnetic compass needle points. Oersted suspected that it was lightning that was responsible for this effect. Franklin had already suggested that lightning was an electrical current, so Oersted reckoned that a wire that glowed from an electrical current passing through it would act like lightning and disturb a magnetic needle. During a private lecture before a group of students in the spring of 1820, he placed a conducting wire over and parallel to a magnetic needle (fig. 7.1) . The compass needle moved, not in the direction of the current but instead in a position at a right angle to the current, as though a magnet had been moved close to it. Oersted reversed the direction of the current. The compass needle swung again, this time in the opposite direction, again at a right angle. The way in which the needle turned, whether to the right or the left of its usual position, depended on the position of the wire that carried the current—whether it was above or below the needle—and on the direction in which the current flowed through the wire.

In demonstrating that the movement of the needle indicated the direction of the current, Oersted had discovered the principle of the galvanometer, or indicator of currents. (The term

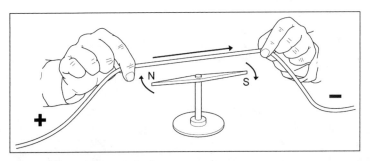

Figure 7.1: Drawing of Oersted's experiment. Artist: Jeff Dixon.

galvanometer is applied to a device that measures the strength of electric currents by means of the deflection of a magnetic needle, around which a current is caused to flow through a coil of wire.) More important, he had demonstrated for the first time in front of his students that a connection exists between electricity and magnetism. Oersted's lecture was one of those exceedingly rare occasions in the annals of science when a major scientific discovery has been made in front of an audience of students during a lecture.

Oersted repeated the experiment later using a more powerful battery and larger conductor. The result was the same: A current passing through a wire could cause the movement of a magnetic compass needle. He was especially taken with the way in which magnetic needle and current interacted, which seemed different from the Newtonian manner of central attraction or repulsion. When a magnetic needle was brought into the field surrounding the wire, it set itself tangent to the circular field. If the needle was carried around the wire, it would continue its tangential position pointing in one direction when beneath the wire and in the opposite direction when above it. If the direction of the current was reversed in the conductor, the direction of the needle was similarly reversed from its former position. Various substances interposed between wire and needle had no effect on the latter.

The scientific community had long championed Coulomb's opinion that electrical and magnetic forces were completely independent forces; the vast majority of scientists regarded the apparent similarities between electric and magnetic forces as no more than an interesting coincidence. As late as 1819, the French physicist Jean-Baptiste Biot (1774–1862), in an article written for the *Edinburgh Encyclopedia*, insisted that magnetism and electricity were independent phenomena. This prevailing opinion was exploded by Oersted's announcement of electromagnetism in 1820.

Scientists were startled by the simplicity of his experiment: The electric current had been a commonplace in physical and electric laboratories for more than 20 years. Why had no one carried out the straightforward experiment Oersted now reported? Other researchers had attempted this experiment, but they had commenced by placing the compass at right angles to the wire and so no effect was observed. Intuitively, they expected that the magnetism created by the current should act in the direction of the current. In that case, the needle should swing parallel to the wire. As it happens, the magnetic force produced by a current is directed at right angles to the current, which means that a compass needle would swing away from the axis of the wire.

Oersted's explanation for his success speaks volumes about how theories can close off avenues of research. He said that no one had looked for a magnetic effect of an electric current because Coulomb's experiments in the 1780s had apparently proved conclusively that electricity and magnetism were two entirely different species of matter, independent of one another. Only someone like Oersted, who had rejected the material theory of the imponderables and believed in the unity of force, would undertake to detect a conversion of one force into another. This discovery was a vindication of his faith that had influenced Oersted's work for many years. The foundation of this faith was the Kantian conviction of the unity of the forces of nature—a philosophical conviction that motivated his search for definitive empirical evidence.

With this simple experiment, the study of electromagnetism was born. It proved to be one of the most fruitful areas of study in nineteenth-century science. Oersted's great discovery was announced in a four-page memoir entitled *"Experimenta circa effectum conflictus electrici in acum magneticum"* (Experiments on the Effects of a Current of Electricity on the Magnetic Needle). It was privately printed and circulated by Oersted to friends and science groups in Europe. He was showered with honors and awards of all kinds. He was elected to fellowship in many learned societies, and the Royal Society awarded him the Copley Medal. The Institut de France presented him with a prize of 3,000 gold francs, a prize similar to that presented to Davy for his discoveries in electrochemistry.

THE ELECTRODYNAMIC MOLECULE

Oersted's paper was published in 1820, and soon electrical experimenters were correlating current flow and magnetism. The news of Oersted's discovery quickly spread to France, where André Marie Ampère (1775–1836), professor of mathematical analysis in the French École Polytechnique (1809) with an interest in theoretical chemistry, witnessed a demonstration of electromagnetism by Arago at the Academy of Sciences on the September 11, 1820—just a week after Oersted's work was reported to the academy in Paris.

It had been known for some time that two magnets attract or repel each other, if free to move. It was also known that attractions and repulsions take place between electrified bodies. Now, Arago demonstrated to an audience of eminent skeptics that a cylindrical coil of wire carrying current could magnetize iron filings, just as a magnet would. It appeared that a magnetic force, indistinguishable from that of ordinary magnets, resided in the electric current—indeed, that a flow of electric current was a magnet. (The iron and magnetizing wire together are called an *electromagnet*, and Arago was the first scientist to build one. More on this in the following section.)

Ampère, who at the time of Arago's demonstration served as permanent secretary of the École Polytechnique, and who later served as professor of physics at the Académie Royale des Sciences, was one of the many scientists who were startled by Oersted's discovery. In step with the majority of scientists, he had maintained that only electricity could interact with electricity, magnetism with magnetism, and so forth. He had even argued in lectures that electrical and magnetic phenomena are due to two different fluids that act independently of each other.

Ampère's scientific work can be divided into four periods. The first was devoted to pure mathematics. The second, from 1808 to 1815, was dominated by studies in chemistry; he worked on halogens and chemical theory. He was a devout atomist and arrived at Avogadro's hypothesis later than Avogadro but independently. The third period, from Arago's demonstration in 1820 to 1827, was devoted to studies in electromagnetism. In the fourth period near the end of his life, he turned to philosophy and to the classification of the sciences.

Following Arago's demonstration, Ampère returned to his own laboratory to work out the consequences of this startling new discovery. What piqued his interest, in particular, was what happened when a compass needle was held over or under a current-carrying wire. Ampère could make no physical sense of the variations in the deviation of the compass needle and its setting at 45 degrees. In point of fact, what Oersted had reported was the *combined* effect of the force from the wire and that from the Earth's magnetic field. Ampère realized that the reported electromagnetic effect would not be isolated until the influence of the Earth's magnetic pull was neutralized, so that the orientation of the needle depended only on the action of the wire.

To this end, he positioned magnets so that their net effect precisely balanced and canceled out the influence of the Earth's magnetic force through the space where the apparatus was mounted. Such an arrangement is now called an *astatic needle*.[1] It allowed Ampère to observe that the compass needle was deflected until it came to rest at right angles to the wire. It was immediately apparent to him that this meant that the magnetic force formed a circle in space, concentric about the wire. Oersted had suspected that this would be the case, but it was not clear from his experiments.

In short order, Ampère set up a series of experiments to determine the exact relationships of current flow and magnetism. His apparatus (fig. 7.2) consisted of a circuit divided in such a way that the portion *CD* was movable, while the portion *AB* was fixed. The ends of the movable framework were dipped into mercury cups by which the current was led away. The whole apparatus was enclosed in a glass container to avoid disturbing currents. As soon as the circuit was complete with the current flowing in the same sense along *AB* and *CD*, the movable portion was attracted.

If one current-carrying wire was accompanied by magnetic force, he reasoned that the magnetic forces of two currentcarrying wires should interact. Within a week of Oersted's announcement, on September 18, he presented a paper entitled "*Mémoire sur l'action mutuelle de deux courants électriques*" (Memoir on the Mutual Action of Two Electric Currents) at the Académie Royale des Sciences in which he demonstrated that two parallel wires carrying current attract each other if the currents are in the same direction and repel each other if they are in the opposite direction. This law is true whether the parallel wires are part of two different circuits or parts of the same circuit. This result, he believed, highlighted an important difference between current electricity and static electricity. In the case of charged bodies, unlike charges attract and like

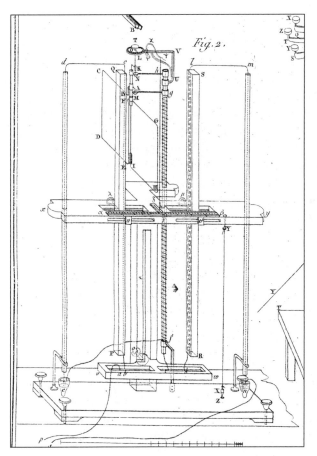

Figure 7.2: Ampère built a number of different apparatuses to study current phenomena. From Ampère, *Annales de Chimie et de Physique* (1820).

charges repel one another, whereas in the case of electric currents, two parallel portions of a circuit are mutually attracted if the currents are in the same direction and repel if the currents are in the opposite direction.

This paper also details Ampère's theory of the magnet, asserting that a magnet can be conceived as an assemblage of minute currents of electricity, whirling all with the same direction of rotation around the steel molecules and in planes at right angles to the axis of the bar. This theory, he claimed, would account for all the known properties of a magnet. To support this view, he produced magnets formed only of spools of copper wire traversed by electrical currents. "What would Newton, Dufay, Aepinus, Franklin, and Coulomb have said if one had told them that the day would come when a navigator would be able to lay the course of his vessel without a magnetic needle and solely by means of electric currents?" (qtd. in Motteley 1922, p. 472). Magnetism, it appeared, was merely electricity in motion.

Coulomb had resisted the possibility of an interaction between electricity and magnetism on the grounds that "like only acted upon like." Ampère's theory of the magnet, in a sense, was true to Coulomb's dictum. The action of an electric current on a magnetic needle was not an interaction between two different kinds of things but rather for Ampère the action of electricity upon itself. The proof was that two current-carrying wires did affect one another.

Ampère's paper of September 18, 1820, was the first of five papers by him that describe the action of conductors carrying electric currents in mutually attracting and repelling one another. The haste with which these papers were produced was not typical of Ampère, whose practice was to revise his work extensively before presenting it to the public. Together these papers laid out the basic physics of the science of *electrodynamics*, the name that he coined for the new science of the needle and the current.

If electricity in motion was magnetism, then in principle it should be possible to reproduce all the effects of permanent magnets by suitably arranged current-carrying wires. Ampère showed that a wire wound into a helix had all the properties of a magnet. In so doing, he provided scientists with a way of fashioning magnetic fields without dependence on lodestones or steel bars, for an easily controlled electric current now provided an even stronger form of magnet. In this way, he rendered untenable the traditional sharp distinction between the fields of magnetic and electrical phenomena.

Ampère fashioned a model to clarify just why an electrical current should circulate around the axis of a magnet. Permanent magnets, he submitted, were simply iron bars in which electric currents flowed in circular currents perpendicular to the axis of the bar. In personal correspondence, Ampère's friend, Augustin Fresnel (1788–1827), who extended the wave theory of light to a wide range of optical phenomena, pointed out that permanent magnets reveal no sign of the presence of electrical currents in them. Nor did a magnet decompose water when it was plunged into a glass of water. Fresnel suggested a different model that Ampère was willing to adopt as his own. Because all that was required were electrical currents whose resultant effect would be a coaxial current, why not simply assume electrical currents around each iron molecule and then envision the process of magnetization as the alignment of these molecules? Because we know nothing about the laws governing the behavior of molecules, electrical currents might circulate around molecules and not be expected to have the side effects accompanying electrical currents on the macroscopic level.

Adopting this concept of an *electrodynamic molecule*, Ampère went on to define an *electrical current* as a flow of positive and negative electricity past one another in opposite directions. In the molecule, the current originated deep in the molecule's interior and then flowed outside the molecule in opposite directions from one pole to the other. This model was the starting point for a number of studies undertaken by Ampère in the years 1821 to 1825, which laid out the mathematics of the new science of electrodynamics.

These results were reported in Ampère's magnum opus—*Mémoire sur la théorie mathématique des phénomènes électrodynamique, uniquement déduite de l'expérience* (1827: Memoir on the Mathematical Theory of Electrodynamic Phenomena Deduced Solely from Experiment). This work, which reduces the laws of electrical action to mathematical form, has been called by some observers "the *Principia* of the science of electrodynamics." The reality of the new electrodynamic molecule of Ampère rested on such an imposing structure of mathematics that only a mathematical illiterate could doubt of its existence.

The discovery that an electrostatic charge magnetized iron had already pointed to a correlation between electricity and magnetism. Now Arago showed that a current could magnetize iron, as well as act upon existing magnets. Into the axis of a galvanic conductor made in the form of a coil, or helix, he placed a needle, the extremities of the wire being connected to the poles of a battery, and with this he showed that the wire not only acted on bodies already magnetized, but that it could develop magnetism in such as did not already possess the power. At first he used soft iron and found that the induced magnetism was only temporary. On repeating his experiment, however, he succeeded in permanently magnetizing small steel needles.

Almost immediately, the new technique was employed to construct electromagnets—combinations of iron and magnetizing wire—far more powerful than any magnets made previously. The great usefulness of the electromagnet in its application is that its magnetism is under the control of the current; when a circuit is made, it becomes a magnet; when the circuit is broken, it ceases to be a magnet.

Ampère was not the only one to react quickly to Arago's report of Oersted's experiment. Biot, with his assistant Felix Savart (1791–1841), also with some haste conducted a series of experiments that led to the formulation of the Biot-Savart law. Reported to the academy in October 1820, this law states that the intensity of the magnetic field set up by a current flowing through a wire varies inversely with the square of the distance from the wire. The Biot-Savart law can be thought of as the magnetic counterpart of Coulomb's law, but it is logically equivalent to Ampère's law. This principle is fundamental to modern electromagnetic theory.

Another French scholar who worked on magnetism at this time was Poisson, who insisted on treating magnetism without any reference to electricity. Poisson had already written two important memoirs on electricity, and he published two on magnetism in 1826.

MEASUREMENT OF CURRENT

Although the 1820s was a time of great discoveries concerning the production of current electricity, it was also a time of widespread confusion about proper definitions for such fundamental terms as *tension, intensity,* and *quantity*. Ampère had provided an electromagnetic definition of the intensity of a current, and had clearly drawn a distinction between electrical tension and electricity, but scientists still lacked a means of relating the tension of a voltaic pile to the intensity of the current it produces and to the properties of the conductor carrying this current. A complete theory of voltaic circuits, which takes into account the driving power of the battery, was first provided in 1826 by the German physicist Georg Simon Ohm (1789–1854).

In 1825, Ohm resolved to dedicate himself to research in electricity. In his research, he was inspired by the work on heat of Joseph Fourier (1768–1830), who pointed out in 1822 that the flow of heat between any two points depends on the temperature difference and the conductivity between them. Applying this insight to the case of electricity, Ohm reasoned in a general way that the flow of current is proportional to the voltage (the electromotive force) and inversely proportional to the resistance.

To support his hunch, Ohm conducted a series of experiments in an attempt to measure the deflection of a magnetic needle held over a wire that was connected between the terminals of a voltaic pile. The guiding principle of the experiment was that current passing through the wire would cause the needle to turn, which could then be measured. The results of these experiments were not altogether satisfactory. The pile itself was decidedly unwieldy, forcing Ohm to contrive mathematically demanding methods for normalizing his data.

A second series of experiments was more fruitful. Ohm first determined the lengths of wires, made from different metals, which gave the same current. He called these various lengths "equivalent lengths." Ohm then showed that the resistance was proportional to the cross-sectional area of the wire and inversely proportional to its length. This result went a long way to explaining results of Davy and other scientists who were working on galvanic electricity.

Ohm then returned to his first series of experiments, replacing the voltaic pile with a thermocouple (fig. 7.3), recently discovered in 1821 by Thomas Johann Seebeck (1770–1831). The thermocouple had the advantage over of the voltaic pile of not being prone to fluctuations in voltage. After trying the experiment with different metals and temperatures, Ohm was able to publish the law that bears his name, which appeared for the first time in his *Die galvanische Kette, mathematisch bearbeitet* (1827: *The Galvanic Circuit Investigated Mathematically*). Using the units adopted by electricians, Ohm's law can be restated as follows: The number of amperes of current (C) flowing through a circuit is equal to the number of volts of electromotive force (E) divided by the number of ohms of resistance (R) in the entire circuit; or $C = E/R$. With this law, scientists could for the first time work out the amounts of current, voltage, and resistance in electric currents and predict how changes in one element would impact on changes in the other elements. In turn, this enabled scientists to design circuits to perform specific functions.

Figure 7.3: Ohm's experimental arrangement. Courtesy of Corbis.

NOTE

1. The precision astatic galvanometer was developed in 1825 by Leopold Nobili (1784–1835), professor of physics at Florence. Using two identical magnetic needles of opposite polarity, he found that he was able to compensate for the effects of the Earth's magnetic field. With this instrument, he managed to detect the flow of current in the body of a frog from muscles to spinal cord. He detected the electricity running along saline-moistened cotton thread joining the dissected frog's legs in one jar to its body in another jar. Nobili was working to support the theory of animal electricity and this conduction, transmitted without wires, he felt demonstrated animal electricity.

8

FORCES AND FIELDS

FARADAY'S ROTATIONS

In the wake of Oersted's announcement of the discovery of electromagnetism, editors of scientific journals were swamped with articles on the phenomenon. Michael Faraday (1791–1867) was also carried away by the hubbub surrounding Oersted's experiment. His initial reaction was to fall back on Davy's position that the motions of the needle resulted from attractions or repulsions of the poles and wire. Faraday's position was changed in 1821 when he was invited by the editor of the *Annals of Philosophy* to write a brief historical survey of the evolution of the science of electromagnetism. Although only eight months had passed since Oersted's communication, so much had already been written that Faraday found it difficult to make sense of the many new theories that had been advanced to account for the phenomenon. He used the occasion to repeat systematically almost all of the previous experiments in electromagnetism and to consider the merits of the rival theories.

Oersted's compass suggested that the magnetic field was not traversing the wire end to end in a straight line but was instead circling the wire. Faraday set up a simple experiment of his own in 1821, showing that a current-carrying wire could be made to move around a fixed magnet, and that a magnet could be made to move around a fixed wire (Faraday 1821b). In Faraday's experiment (fig. 8.1), a circuit consisting of two vessels holding mercury and connecting wires was arranged so that one vessel held a fixed wire and a movable magnet and the other held a fixed magnet and a movable wire. The current passed from the wire through the mercury in the first vessel to a copper pin running into the base of the vessel. The magnet in this vessel was fastened to the copper pin by a thread. In the second vessel, the fixed magnet was placed in a socket in the

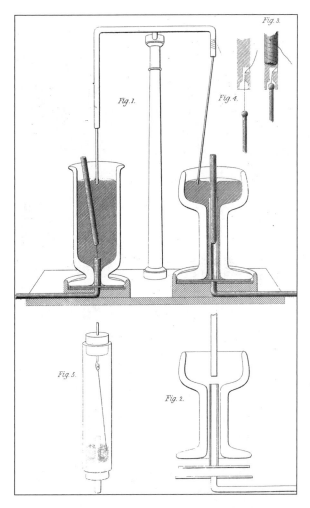

Figure 8.1: Experimental apparatus for Faraday's rotations. From Faraday, *Experimental Researches in Electricity*, vol. 2 (1844).

stem, and the movable wire, dipping into the mercury, was able to move freely by means of a ball-and-socket connection.

Faraday's rotations accommodated the various attractions and repulsions described by Oersted, Davy, and Ampère. Indeed, the action of any wire system on a magnetic pole could be traced to the combined circular effects of the different portions of the wire. The announcement of Faraday's discovery in "On some new Electro-Magnetical Motions, and on the Theory of Magnetism" was made in the October 1821 issue of the *Quarterly Journal of Science*.

Faraday's electromagnetic rotations quickly swept Europe as a new and important discovery, establishing his reputation as one of the leading scientists of the day. The rotations were surprising because they were the first example of a device that, seemingly on its own, could start up and continue in motion constantly, against friction. The battery was depleted gradually, so this was no perpetual-motion machine. Using a broad definition of *motor* as meaning any device that converts electrical energy into motion, Faraday's apparatus can be construed as the first electric motor. Although useful as demonstration devices, the fact remains that Faraday's moveable magnet and wheel do not resemble anything that most people today would recognize as an electric motor.

Faraday sent a small kit of his rotation device, containing the cork, the wire, and the glass tube, to Ampère in September 1821. To repeat the experiment, all that was required of Ampère was that he pour mercury into the tube and connect it to a galvanic battery. Although Faraday's continuous rotations were implicit in Ampère's theory, it is instructive to consider why Ampère did not discover them. We have already seen that theories sometimes effectively close off potentially fruitful avenues of investigation. The most obvious reason is that Ampère's electromagnetic theory was wedded to the analogy with gravitational attraction with its central forces that acted in straight lines between the origins of the forces; this analogy with Newton's central forces effectively masked the phenomenon of rotational motion. As one might expect, Ampère's considered judgment was that Faraday's "circular force" was the resultant of all the central forces generated by the electrical current that act directly in a straight line between the charged particles or magnetic poles. The primary fact for Ampère, then, was the poles or current elements in the wire from which the electromagnetic forces emanated. When summed vectorially, he claimed that these forces produced the circular force.

A less obvious but equally salient factor in Ampère's failure to discover continuous rotations was the character of his experimental apparatus—a refined device that was built by a professional as a means of proving consequences of his theory that he could not predict. As a result, Ampère's apparatus had a rigid character, in stark contrast with Faraday's handcrafted devices of wires and needles, which could be modified with ease, thereby allowing Faraday to pursue unexpected effects. Given the rigidity of his devices and their role in verifying theoretical claims, Ampère was reluctant to modify his apparatus in the spirit of exploring unexpected effects. When Ampère produced a current by electromagnetic induction in an experiment of 1822 (Hofmann 1987), it is not surprising that he did not pursue this matter and thereby anticipate Faraday's discovery of electromagnetic induction eight years later.

Faraday could not follow Ampère's intricate mathematical calculations and was nevertheless unimpressed by them. What mattered to him was that he could create and control the production of these circular rotations in his laboratory. As fair turnabout, Faraday designed an ingenious experiment to show that Ampère's prized central forces could be the resultant of a circular force. He wound a helix around a glass tube and suspended it half-submerged in a pan of water. He then took a magnetic needle as long as the helix and fixed it to a cork so that it would float. When the current was turned on in the helix, the north pole of the floating needle moved toward the south pole of the helix. If this were a simple case of the north pole of the helix attracting the south pole of the magnet, Faraday reasoned, when the two met, they should cling together with some force. This did not happen. The needle entered the helix and the north pole of the needle continued to traverse the tube until it reached the north pole of the helix. At this point, it came to rest. Faraday concluded that the motion of the needle showed that the line continued in the magnet and did not terminate at the poles.

POWERS

Faraday's work occasioned a nasty row with his mentor, Humphry Davy, who claimed that Faraday had been led to his experiment by a conversation that he overheard between Davy and William Wollaston (1766–1828), the

English chemist and physicist who discovered the elements palladium and rhodium and first reported the dark lines in the spectrum of the Sun. This allegation proved to be most difficult for Faraday. After we look at the relationship between Faraday and Davy, we will return to this conversation and interrogate this allegation of intellectual theft in some detail.

Faraday was born to a poor and very religious family in what is now South London—his father a blacksmith and his mother a country woman who supported her son emotionally through a difficult childhood. Faraday was one of four children, all of whom were hard put to get enough to eat because their father was often ill and incapable of working steadily. The Faradays belonged to the Sandemanians—a small Christian sect dedicated to the idea that all persons have the capacity to find the truth for themselves, obviating the need for clerics to serve as interpreters of the Bible. Shortly after his marriage, at the age of 30, Faraday joined the same sect, to which he adhered until his death. Religion and science he kept strictly apart, believing that the data of science were of an entirely different nature from the direct communications between God and the soul on which his religious faith was based.

Faraday received only the fundamentals of an education, learning to read, write, and cipher in a church Sunday school. At the age of 14, he was apprenticed to a bookbinder. He took this vocation as an opportunity to read some of the books brought in for rebinding. An article on the subject of electricity in the third edition of the *Encyclopædia Britannica* particularly fascinated him. Using old bottles and lumber, he made a crude electrostatic generator and did simple experiments. He also built a weak voltaic pile with which he performed experiments in electrochemistry.

Faraday's interest in science was boosted when he was offered a free ticket to attend chemical lectures by Davy at the Royal Institution. Faraday recorded the lectures in his notes, and returned to bookbinding with the seemingly unrealizable hope of entering the temple of science. He decided to send a bound copy of his notes to Davy along with a letter asking for employment, but there was no opening. Davy did not forget, however, and when one of his laboratory assistants was dismissed for brawling, he offered Faraday a job as his assistant at the Royal Institution in March 1813, at a time when Davy had been temporarily blinded by an explosion involving a notoriously unstable substance—chloride of nitrogen.

When Faraday walked through the door of the Royal Institution in 1813, Davy was in the process of revolutionizing the chemistry of the day. Lavoisier, the French scientist generally credited with founding modern chemistry, had effected his rearrangement of chemical knowledge in the 1770s and 1780s by insisting upon a few guiding principles. One of these principles was that oxygen was a unique element, in that it was the only supporter of combustion and was also the element that lay at the basis of all acids. Davy, after having discovered sodium and potassium by using a powerful current from a galvanic battery to decompose oxides of these elements, turned to the decomposition of muriatic (hydrochloric) acid, one of the strongest acids known. The products of the decomposition were hydrogen and a green gas that supported combustion and that, when combined with water, produced an acid. Davy concluded that this gas was an element, to which he gave the name chlorine. He also was quick to note that there was no oxygen whatsoever in muriatic acid. Acidity, therefore, was not the result of the presence of an acid-forming element but of some other condition. What else could that condition be but the physical form of the acid molecule itself? Davy suggested, then, that chemical properties were determined not by specific elements alone but also by the ways in which these elements were arranged in molecules.

Under Davy's tutelage, Faraday soon became an outstanding chemist (fig. 8.2). Davy made amends for Faraday's lacking a formal education by taking him along on a trip to Europe, where he was introduced to the key scientific figures of his day, including Volta, Ampère, Joseph-Louis Gay-Lussac (1778–1850), Arago, Alexander von Humboldt (1769–1859), and Georges Cuvier (1769–1832). It has been said that finding Michael Faraday was Davy's most important finding—a comment that is insensitive to the importance of Davy's contributions to chemistry and to the fact that it was Faraday who found Davy.

In recognition of his discovery of the continuous rotations, in 1824 Faraday was made a Fellow of the Royal Society with Wollaston as its president, the lone dissenting vote being cast by Davy. This failed blackball, and the suggestion that he had appropriated the ideas of others, were not damaging to Faraday's reputation in the long run, however, because his demonstration of continuous rotations was only the first of his many important discoveries.

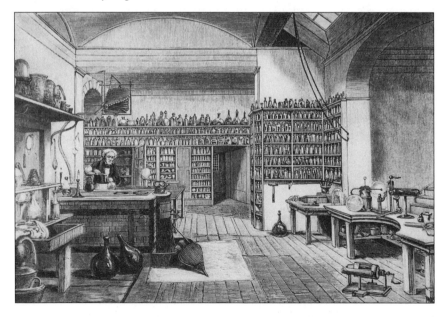

Figure 8.2: Faraday in his physics laboratory at the Royal Institution. Courtesy of Corbis.

When we interrogate the circumstances surrounding Davy's conversation with Wollaston, it is plain that the allegation of impropriety was groundless. Wollaston's aptitude for experiment was considerable. Reference has already been made to the importance of his contributions to chemistry, and his work in electricity and magnetism was of the same order. Many years of experience alerted him to the clash between Newtonian central forces and Oersted's electromagnetic effect. Circular forces, Wollaston was convinced, would explain this action. To render this hypothesis plausible, he designed an experiment that involved placing a permanent magnet in the immediate vicinity of a current-carrying wire. The electromagnetic effect, he was convinced, was the result of a helical current rotating around its own axis.

With Davy's assistance, Wollaston carried out this experiment in the laboratory of the Royal Institution in April 1821. The experiment was a resounding failure; they were unable to produce rotations of electromagnetic currents. Faraday was not present at these experiments. When he did arrive at the laboratory and proceeded to go about his usual laboratory business, Davy and Wollaston were discussing these experiments. There is no evidence that the discussion made any great impression on Faraday. If anything, Faraday's laboratory diary testifies that he was consumed at the time by laborious experiments on alloys of steel. Faraday conceded that he may have gotten a start from the conversation between Davy and Wollaston, but nothing more.

Once rumors began to circulate that Faraday had applied the content of the conversation to his own personal benefit, Faraday invited Wollaston to his laboratory to observe several experiments. What these experiments showed conclusively was the originality of his apparatus, a point subsequently (though not enthusiastically) corroborated by Wollaston. They also demonstrated that the interaction between the magnetic needle and the current-carrying wire was not as simple as Wollaston had assumed. In his *Diary* (vol. 1:50), Faraday wrote: "magnets of different power brought perpendicularly to this wire did not make it revolve as Dr. Wollaston expected, but thrust it from side to side." Faraday's simple apparatus demonstrated how this passing off at a right angle from the pole could be translated into an actual rotation. After observing Faraday and his experiments, Wollaston was convinced that his idea had not been appropriated and the rumors ceased.

ELECTRICITY FROM MAGNETISM

The period from 1821 to 1831 was one of intense activity for Faraday, and there was precious little time for the study of electricity and magnetism. He did some work in chemistry, liquefying chlorine in 1823 and discovering benzene in 1825. In 1825, he was appointed to the post of director of the laboratory and professor of chemistry at the Royal Institution. In the late 1820s he undertook an extensive project on making optical glass for a joint committee of the Royal Society and Board of Longitude. He was always in demand for lectures, and his Friday Evening Discourses commenced in 1826, continuing to 1862. The year 1826 also marked the commencement of his Christmas Lectures for children, which continued until 1861. Although Faraday made occasional forays into the study of electromagnetism, it was not until 1831 that he was able to pursue these studies in a concerted manner.

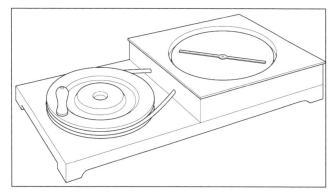

Figure 8.3: Illustration of Arago's disk. Artist: Jeff Dixon.

The induction of currents was a critical, unresolved issue: If a current could induce a magnetic field, it seemed clear to scientists that a magnetic field had to be able to induce a current. Furthermore, the power of static electricity to induce an opposite electrical state on bodies in its vicinity was well known, but no one had been able to show that electrical currents possessed a similar property, despite obvious analogies between voltaic electricity and static electricity.

In 1824, Arago designed an experiment known as *Arago's disk* (fig. 8.3), which showed that a needle suspended on a pivot held above a spinning copper disk will follow the rotation of the copper disk. What was impressive about Arago's apparatus was its utilization of copper, a nonmagnetic substance; that is, the experiment seemed to show that mere motion could induce magnetism in a metal object. Arago's copper disk caused a commotion. Arago found that the other metals exhibited similar effects. Silver showed a greater effect than copper, whereas lead, mercury, and bismuth were less effective. Ampère also worked on this problem during the summer of 1826 and finally replicated Arago's experiment after substituting a solenoid (a coil of wire around a metallic core) for the magnet of Arago's original experiment. Poisson published a theoretic memoir on the subject, but no cause could be assigned for so extraordinary an action.

In 1825, during one of his brief forays into the study of electromagnetism, Faraday tried a number of geometrical configurations of conductors, but his galvanometer was too insensitive to show a disturbance. He tried induction from coil to coil and induction from moving magnet to spiral, but without success. Toward the end of 1831, a new approach occurred to Faraday. His friend, Gerritt Moll (1785–1838) in Holland had noticed in August of that year the almost instantaneous reversal of the polarity of an electromagnet upon reversal of the current passing through it. Moll's work also drew attention to the high intensity of magnetic force that could be produced by a large electromagnet. Faraday considered carefully what shape an electromagnet should have in order for it to be very strong and to act powerfully on a nearby circuit when suddenly magnetized, settling on a thick ring, six inches in diameter and

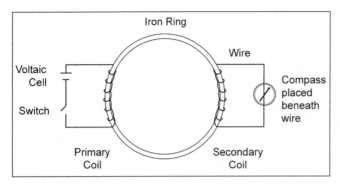

Figure 8.4: Faraday's apparatus for electromagnetic induction. Artist: Jeff Dixon.

seven-eighths of an inch thick cast in iron. He wound on it two sets of coils, one on each half of the ring. The coils were separated by twine and calico. One was connected to a galvanometer, the other to a battery (fig. 8.4).

Faraday thought that the current in the first coil of wire might cause a current in the second coil. The galvanometer would measure the presence of the second current and tell the story. The experimental design was sound (it was the first *transformer*, which is now universally used to convert low voltages to high and vice versa), but it contained a surprise: Despite the steady magnetic force set up in the iron ring, no steady electric current ran through the second coil. Instead, a flash of current ran through the second coil when Faraday closed the circuit, signaled by a feeble jump on the galvanometer. When he opened the circuit again, there was another flash of current, signaled by a second feeble galvanometer jump. Faraday made many variations of this experiment and concluded that the battery current through the one wire in reality did induce a similar current through the other wire; but that it continued for an instant only, and partook more of the nature of the electric wave from a Leyden jar than of the current from a voltaic battery.

During one of his enormously popular lectures at the Royal Institution in 1831, Faraday took a coil of wire and moved a permanent magnet into the coil. The needle of the galvanometer attached to the wire swung, then stopped, when the movement of the magnet stopped. When he moved the magnet out of the coil, the galvanometer registered once more. If he moved the coil of wire over the magnet, a current was again detected. If he let the magnet just sit motionless inside the coil of wire, the galvanometer did not register a current.

Electrical currents were produced in a closed conducting circuit, Faraday found, not by the mere presence of a magnetic field (i.e., winding a conducting wire around a magnet did not produce an electric current) but by changes in the strengths of that field (in proportion to the rate of change of the field). A loop of copper lying near the poles of a horseshoe magnet, both being at rest, will have no electricity generated in it, although it is immersed in a magnetic field. However, if the loop of wire is moved in and out of the magnetic field, or if the magnet is moved to the same purpose, then electricity will be generated in the wire; the faster the motion, the more electricity will be generated in exact proportion. All that was required for the production of current was the relative motion between coil and magnet. It had been difficult to see this in Oersted's experiment because the magnetic needle stayed in a steady position as long as the current in the wire remained steady.

Faraday's discovery immediately lifted the mist surrounding Arago's wheel. When a metallic disk was rotated beneath a freely suspended magnetic needle, currents were induced in it. These currents, in turn, created a magnetic field, which acted upon the needle. Faraday soon discovered that another way of inducing the current was to move the electric conductor while the magnetic source stood still. This was the principle behind his *disk dynamo*, which featured a conducting disk spinning in a magnetic field. The electric circuit was then completed by stationary wires touching the disk on its rim and on its axle, shown on the right side of figure 8.5. This device, which is the simplest illustration of dynamo action, is not a practical generator of electricity (unless one seeks to generate huge currents at very low voltages). To carry such currents, massive conductors would be needed, and the copper wiring used in conventional machinery would be completely inadequate.

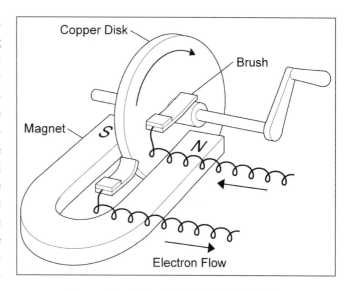

Figure 8.5: Illustration of Faraday's disk. Artist: Jeff Dixon.

Faraday's moving conductor of electricity was solid (e.g., a copper disk), but a circulating fluid can also create such currents. Faraday was aware of the possibility of such "fluid dynamos," and accordingly he tried to measure the induced flow of electricity created by the flow of London's river Thames across the Earth's magnetic field by stretching a wire across London's Waterloo Bridge and dipping its ends into the river. Small voltages due to chemical processes prevented him from observing the effect, but the idea was sound, as evidenced by its central place in modern dynamo theory.

Although the chemical production of electricity was a significant achievement, the cost of batteries made batteries impractical as a source of light and power. What was needed was a device for converting mechanical energy into electrical energy, and the principle of the dynamo became the cornerstone of the electric industry that would soon flourish. It is the principle on which almost all modern electrical machines are built; few discoveries have had a more far-reaching effect. History now credits Faraday with this discovery, despite the fact that the American inventor Joseph Henry (1797–1878) discovered this principle of electromagnetic induction in 1830. However, Faraday made the same discovery independently and published first in a paper that was read before the Royal Society on November 24, 1831.

Faraday fashioned a phenomenological model to explain his discovery of electromagnetic induction. The induction of an electric current between two coils of wire wound round an iron ring, Faraday reasoned, set up a special condition or state in the iron ring, whereby the *particles* of the ring were in a state of tension. It was this so-called electrotonic state that caused the induction of the current in the secondary circuit. When the current ceased to flow in the primary circuit, the electrotonic state collapsed (i.e., the tension was removed), causing the deflection of the galvanometer in the opposite direction.

The electrotonic state was a hypothetical state into which all conducting substances were thrown when subject to magnetic influence. It was completely undetectable, but Faraday the experimenter par excellence clung to this idea for more than 20 years. His attachment to this hypothetical entity might seem out of character, but it was his only option to the old theory of fluids, which he regarded as completely without foundation.

Although the electrotonic state was completely undetectable, Faraday entertained some fairly precise ideas as to its nature. For Faraday, the central feature of the electrotonic state was its association with particles of matter. Faraday was not proposing an atomic theory per se; that is, the term *particle* did not carry the Daltonian connotation of indivisibility. Although Faraday's talk of particles was undeniably atomistic, he did not believe that his phenomenological model called for the development of a fully fledged atomic theory of matter. Nevertheless, in advancing this theory of electrical action as a tension of particles, Faraday was embracing a particulate theory of matter that was to dominate his thinking until 1844 or so.

Another feature of this model, which we have touched on briefly, is its *provisional* character. Although his many experiments were organized by such theoretical notions as the "electrotonic state," Faraday's phenomenological models were easily revisable. Indeed, Faraday was proud of the flexibility of his conceptual scheme, in which the nature of the electric current remained undetermined (Faraday 1839, 159).

EXPERIMENTS ON ELECTROLYTIC DECOMPOSITION

Faraday's first and second series of experimental researches were concerned with the relationship between electricity and magnetism. Continuing in the path of Volta and his former mentor, Humphrey Davy, in the summer of 1832, Faraday turned to the question of the relationship between different manifestations of electricity. His commitment to the doctrine of the convertibility of forces led him to believe that the electricities produced by traditional electrostatic machines, voltaic cells, thermocouples, dynamos, and electric fishes were forms of the same force. Faraday's conviction that all these manifestations were forms of the same force was challenged in a letter to the *Philosophical Transactions* by the late Humphry Davy's brother, John Davy (1790–1868), who insisted that electrical effects were not produced by a single agent but were the complex results of a combination of powers.

Faraday's first order of business in his attempt to establish the identity of the electricities was identifying the effects produced by electricity in its various forms. He listed six such effects: the attraction and repulsions of static charges, the evolution of heat, magnetism, chemical decomposition, physiological effects, and the spark. He then examined voltaic electricity, static electricity, magnetoelectricity, thermoelectricity, and animal electricity to ascertain how far they could be made to exhibit these effects. Most of the electricities were shown to be like other forms of electricity by the citation of articles. The sticking point was static electricity. With respect to chemical decomposition, there were two obvious differences in the behavior of voltaic and static electricity. When Wollaston attempted to decompose water in 1801 by discharging a spark from an electrostatic generator through a solution, the water was decomposed, but hydrogen and oxygen bubbled up at both wires. Voltaic decomposition, in contrast, resulted in hydrogen appearing at one pole and oxygen at the other. The second difference was the magnetic effect of static electricity. Although the power of electrostatic discharges to magnetize needles was evident to all, the deflection of magnetic needles was still in question.

Faraday attacked these worries with a series of original experiments that resulted in his empirical generalization known as Faraday's laws of electrolysis. He elected to focus on the magnetic effect of static electricity, deciding that the simplest way to study this relationship was to record the deflection of a galvanometer needle by discharge. He soon discovered that the discharge itself affected the magnetism of the galvanometer. Faraday hit upon a way of eliminating this disturbance. Passing the electrostatic charge through a wet string, which was a poor conductor, he was able to slow down its passage, permitting the electrostatic discharge to act more like a current (fig. 8.6). He then spread out the solution to be decomposed on a glass plate to minimize the volume effects. When the electrostatic discharge was passed through it, true electrochemical decomposition took place. This experiment left little doubt as to the identity of static and voltaic electricity.

Because the galvanometer could be deflected by static discharge, Faraday discerned an opportunity to compare the quantities of electricity that passed in voltaic circuits and electrostatic discharge. By passing equal quantities of electricity (static and voltaic) through water, he discovered that the same amount of electrochemical decomposition took place.

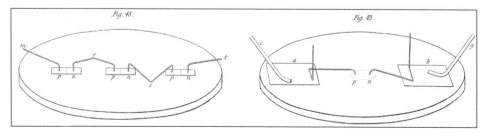

Figure 8.6: Faraday's apparatus for electrolytic decomposition. From Faraday, *Experimental Researches in Electricity*, vol. 1 (1839).

Faraday was far from finished. To ensure that the phenomenon was genuine, he made a subtle change to his apparatus. In his revamped experimental setup, there were no poles as such but merely a path along which the electrostatic charge could pass. Electrochemical decomposition still took place. It seemed clear that all that was necessary to explain the decomposition was the passage of electricity.

According to the received theory, the poles of the voltaic cell were the centers of attractive and repulsive forces, which literally tore the molecules of a solution into pieces. Since the invention of the Voltaic pile, electrochemists had assumed that the plus and minus terminals of the circuit acted as centers of force. Chemical decomposition occurred when these centers of force, acting at a distance, literally ripped apart the molecules in the electrolytic solution. The action was the familiar Newtonian one of action at a distance, the fragments migrating to their respective poles under the influence of the attractive and repulsive forces. Faraday's result indicated, in contrast, that electricity could not be a material substance that acted at a distance upon the molecules in the solution.

These experiments led directly to Faraday's two laws of electrochemistry:

1. The amount of a substance deposited on each electrode of an electrolytic cell is directly proportional to the quantity of electricity passed through the cell.
2. The quantities of different elements deposited by a given amount of electricity are in the ratio of their chemical equivalent weights.

Experiments on electrolytic decomposition also led Faraday to reject the familiar electrochemical terminology. Phrases like "positive pole" were saturated with the weary fluid concept of electricity and referred to electrostatic action at a distance. Faraday elected to replace the standard terminology with neutral terms "without involving any theory of the nature of electricity" (Faraday 1838, 519–520). He wrote for advice to William Whewell (1794–1866), master of Trinity College at Cambridge. In collaboration with this philosopher, mathematician, and inventor of the new term *scientist*, they concocted a new vocabulary for electrochemical decomposition: *electrode* (electron, electricity, *odos*, a way), *anode* (*ana*, upwards), and *cathode* (*kata*, downwards), for the terminations of the decomposing matter; *electrolysis* for the decomposition itself; *electrolyte* (*luo*, to set free) for a substance decomposed by the current; *ion* (ion, going), *anion* and *cation* for the electrified substances that were transported through liquids in electrolysis. These are all now household words in electrochemistry.

Sometime later in 1881, the German physicist Hermann von Helmholtz (1821–1894) in his Faraday Memorial Lecture, asserted that Faraday's second law of electrochemistry clearly pointed to matter having an atomic nature. If elements were distinguished from one another solely by their atomic weight, and if the same amount of electricity deposited amounts of these elements upon the poles in ratios of their equivalent weights, then it would appear that on the atomic level each element received the same unit, or quantum, of electricity. Today we would call this quantum the electron.

Faraday was far from embracing an atomic theory of matter, but his findings led him to formulate a new theory of electrochemistry. The electric force, he argued, threw the m*olecules* (a term he now substituted for the term *particle*) of a solution into a state of tension (the electrotonic state). When the force was strong enough to distort the fields of forces that held the molecules together and thus to permit the interaction of these fields with neighboring particles, the tension was relieved by the migration of particles along the lines of tension, the different species of atoms migrating in opposite directions. The amount of electricity that passed, then, was clearly related to the chemical affinities of the substances in solution.

The question for Faraday's account of electrostatic induction was how electric tension was transmitted from one particle to another. Once matter was subjected to electrical tension, this led to a state of polarity in the molecules—by polarity, he meant "a disposition of force by which the same molecule acquires opposite powers on different parts" (Faraday 1838, 411). Again, "induction appears to consist in a certain polarized state of the particles, into which they are thrown by the electrified body sustaining the action, the particles assuming positive and negative points or parts, which are symmetrically arranged with respect to each other" (Faraday 1838, 409). The question, however, was how the electric tension was transmitted from one polarized particle to another.

Faraday held fast to the claim that all particles, whether of insulators or of conductors, were capable of being polarized by the influence of neighboring charged particles. When forces were communicated between contiguous particles, induction occurred; when communication of the forces was difficult, insulation occurred. The difficulty for Faraday's theory was just what he meant by the notion of "contiguity." By contiguous particles, he did not mean particles that actually touch one another. What he meant was that they were next to each other, not that there was no space between them (1838, 530-31). It was not part of his view that the particles become polarized due to mutual contact. "It does not follow from this theory," Faraday admitted, "that the particles on opposite sides of … a vacuum could not act on each other … nothing in my present view forbids that the particle should act at the distance of half an inch …" (1838, 514). The problem here was that if action at a distance between contiguous particles was denied, it did not help to suppose that the distance between contiguous particles was insensible. Short-range forces, whose mode of action was undefined, would still have to be supposed as acting at a distance.

Faraday believed that this model could account for all the phenomena of electricity without recourse to electrical fluids. He believed that a physics of force and its interactions with the forces of the material particles could explain magnetism as well. If electricity was a force and was convertible into magnetism, then magnetism too was a force and not a fluid. But magnetism was a peculiar force, restricted to iron, nickel, and cobalt. Electricity appeared to be a universal force, and whatever it was converted into should be universal as well.

FARADAY AND THE THEORY OF MATTER

While the focus of the science of electricity and magnetism was traditionally on the phenomena exhibited by charged and magnetic bodies in and of themselves, Faraday's work on electrolytic decomposition marked a profound shift as he became increasingly concerned with the relations between electricity and matter. Once the properties of electricity and magnetism were established with precision, this shift in focus was inevitable as Faraday and scientists following his lead were at last in a position to exploit the knowledge gleaned about electrical and magnetic properties of substances to learn something about other properties. The established science of electricity and magnetism became, in turn, the foundation for a successor science that increasingly focused on matter itself.

This shift in emphasis is evident in the question that increasingly occupied Faraday's attention—whether electricity has an actual and independent existence as a fluid (or fluids) or was a mere power of matter like what we conceive of as the attraction of gravitation. This question had always occupied Faraday to some extent, but in the mid-1830s it began to consume his other interests. This question involved the crux of the difference between his theory of electrical action and that held by the majority of his peers. All the classical theories—associated with Coulomb, Poisson, Ampère, and others—assumed that electricity could be accounted for by the postulation of fluids distinct from matter. It was this assumption that made investigations into the nature of matter secondary in importance. Faraday rejected the fluids and was convinced that electrical powers were bound up with matter itself. Electricity, in Faraday's view, was inseparably connected to matter.

At the same time, Faraday's heavy workload and relentless intellectual efforts finally began to take a toll. In 1839, he suffered a massive nervous breakdown, marked by memory loss and giddiness. This was followed two years later by the onset of a physical disability involving the use of his legs. It was not until 1845 or so that Faraday had recovered sufficiently to resume work. At this time, he received a letter from the physicist William Thomson (Lord Kelvin, 1824–1907), who suggested that a magnetic field should be able to influence polarized light passing through glass.

Twenty years earlier, Faraday had investigated whether a static magnetic field force might cause some unusual interaction when passed through glass. Thomson's communication prompted Faraday to return to this subject. Substituting a far more powerful electromagnet for a static magnetic field, he discovered that if a ray of polarized light was passed through a transparent material in the presence of a magnetic field, the plane of polarization was rotated. The direction of rotation depended solely upon the polarity of the magnetic line of force; the amount of the rotation was a function of the strength of the magnetic field. The effect was small and only occurred with flint glass, but Faraday's skills as an experimenter unambiguously identified the phenomenon now known as *Faraday rotation*.

This effect convinced Faraday of the universality of magnetism, and he set out to prove that all bodies were affected by magnetic fields. He began by classifying materials into two groups: paramagnetic and diamagnetic materials. The former included materials that were naturally magnetic (magnetite) or could be easily magnetized (platinum). The latter included those materials that conducted the magnetic field poorly (salts of various sorts).

This classification did not sit well with traditional accounts of magnetic excitation, all of which were based on the assumption of the existence of polar molecules in magnetic materials that responded to the presence of magnetic force, in the same way that magnets responded to the Earth's magnetic field. An explanation of diamagnetism that was erected on this assumption would be compelled to resort to a reverse polarity of a sort. Faraday's approach was characteristically different. He stated that diamagnetic materials had no poles. The difference between para- and diamagnetic materials was that magnetic lines of force penetrated the former materials but were unable to penetrate the latter materials. Furthermore, the lines of force themselves have no beginning or end. They do not originate at one pole and travel to the other. Their existence, therefore, could not be assigned to the existence of molecules, each with their own poles, within the magnetized material.

The heart of Faraday's account of magnetism was the idea that the magnetic field is independent of the magnetic material and that the energy was to be found in the field and not in the magnet itself. Magnetism was an interaction of matter with a property in its immediate vicinity. In a lecture delivered at the Royal Institution in 1844, Faraday started to work out a scheme for replacing the theory of action propagated by the polarization of the particles of matter with a theory of lines of force. He invited his audience to imagine the Sun sitting by itself in space, and to envision the Earth suddenly placed alongside it, at its correct distance. He then asked the audience to consider how the Earth would know that the Sun was there. According to Faraday, a web of lines of gravitational force extended outward from the Sun into the universe, and would do so even if there were no planets to feel the Sun's gravitational influence. If the Earth were dropped into the field of force, it would respond instantly to the field at that location. It would not have to wait for some message to travel to the Sun and back to find out that the Sun was actually there and adjust its orbit accordingly.

In his "On the Physical Character of the Lines of Force" (1852), he argued that the lines of force represented a physical state, though their nature remained unclear. Though he did speculate that the lines of force were transmitted by an action that was a "function of the aether" (1855, 3: par. 3075), Faraday was skeptical about the existence of the ether.

APPENDIX: ELECTRIC LIGHTING AND POWER

The social impact of Faraday's discoveries was extraordinary, revolutionizing both the home and the workplace. One role played by electricity was that of transporting energy. In power stations, energy (from the burning of coal, for example) was transformed into electricity, and then transported over large distances to an electrical station, where it was again converted into other forms of energy. Faraday's dynamo transformed mechanical energy into electric current, and the electric motor, which was developed in 1831 by Faraday and by the American inventor Joseph Henry (1797–1878), turned it back into mechanical energy. The first power plants were opened in 1882, and the large plant built at Niagara Falls was opened in 1896. The earliest plants were mainly suppliers of current for electric lighting, but by 1883 the possibility of using electricity for heated pans, cushions, and blankets, carpet cleaners, and a host of other appliances had been recognized.

Lighting was perhaps the most important early application of electricity. In the 1880s, the incandescent lamp replaced the electric arc of pioneer days. It had been known since 1820 that electric currents can make a coil of platinum glow so as to give light. Platinum was a precious metal, and so scientists set out to identify a cheaper material. The solution was found by Thomas Edison (1847–1931) and Joseph Wilson Swan (1828–1914), who invented the incandescent-filament electric lamp. Their joint company, the Edison and Swan United Electric Light Company, came into existence in 1883.

Most people think of Thomas Edison as the inventor of the light bulb, but that is not exactly true. Incandescent lighting—where an electrical current causes a filament to glow—had already been invented. The problem was that the light lasted only a few minutes because the filaments burned out. Edison found a way around that. In 1879 he came up with a workable filament that allowed light bulbs to burn much longer.

A light bulb with no electrical source was not very serviceable, however. Edison knew that a system of generating and distributing electrical power was needed, and he set about creating one. In 1882, Edison opened the Pearl Street Station in New York City. It was the world's first commercial power plant. Although it was an enormous plant for its day, the station at Pearl Street was able to produce and distribute electricity to only one square mile of lower Manhattan.

Although it was successful, Edison's system had problems and limitations. It used direct current (DC), which was not ideal for delivering electricity over long distances. Alternating current (AC) was better suited to the job. AC allows high-voltage current to be transmitted through power lines and lowered before the customer gets it.

Edison was committed to DC. Others, however, were determined to compete with Edison by using AC. AC worked fine with light bulbs, but it was not yet practical for powering electric motors, so the first step was to develop a good AC motor. Finding a way to produce a rotating magnetic field electrically, Nikola Tesla (1856–1943) invented a practical AC motor in 1883. Several years later George Westinghouse (1846-1914) bought the patents to the Tesla motor and set up an AC power system. For a while, Westinghouse and Edison were in a fierce competition to see whose system would dominate. The winner was clear after a new AC power plant opened at Niagara Falls in 1895 and was used to light up portions of Buffalo, New York. Its successful use in transmitting large amounts of power to distant places was the deciding factor. Today, the whole world continues to use AC.

The generation of large amounts of electricity and the development of reliable delivery systems changed the world. Once electricity was cheap and reliable, other inventions followed. Availability of electricity led to widespread use of electric appliances, some of which we still use today. In 1908 the vacuum cleaner came on the market. Five years later an appropriate motor (one that did not require lubrication and was small in size) made the electric refrigerator possible. Washing machines and dishwashers soon came on the scene. In 1935 the clothes dryer was invented. Electricity eventually powered radios, televisions, musical instruments, computers, and countless other devices. Electricity was everywhere and it was changing the way people lived and (with the production of electric chairs) died.

9

THE SCIENCE OF ELECTROMAGNETISM

INTRODUCTION

James Clerk Maxwell (1831–1879) is associated with a number of contributions to nineteenth-century science. He discovered the "fish-eye" lens, determined the nature of Saturn's rings, explored statistics and the physics of molecules, and dabbled in engineering. In this chapter, we will discuss his greatest work, the formulation of a series of equations that summarized all there was to know at the time about electrical and magnetic science. These equations combined the laws of electricity and magnetism with the laws of the behavior of light—areas of physics that scientists had regarded as being distinct from one another. Phenomena that previously were thought to be distinct were shown to be aspects of the same phenomenon. As a result of his synthesis, he helped to unravel some of the mysteries surrounding the properties of light.

GEOMETRICAL IMAGERY AND THE ELECTROMAGNETIC FIELD

Maxwell was born in Edinburgh into a modestly wealthy Scottish family the year of the discovery of electromagnetic induction. As a 14-year-old, he authored his first publication: a paper describing a simple mechanical means of drawing mathematical curves with a piece of string. This combination of algebraic mathematics with elements of geometry would remain a distinctive feature of Maxwell's work.

At age 16, Maxwell left the academy for the University of Edinburgh. This was followed by a movement to Cambridge in 1850, where Maxwell eventually settled at Trinity, a college with a good reputation for mathematics. He graduated in 1854 as second wrangler (i.e., second-highest in the mathematics exam) and co-winner of the highly prestigious Smith Prize in mathematics.

After graduating from the University of Cambridge in 1854, Maxwell took up the study of two subjects—color theory and electricity, the latter of which he found to be more complex than any other science. He was only twenty-three years old at the time. The most comprehensive collection of writings on the subject known to Maxwell was Faraday's three-volume *Experimental Researches in Electricity* (1839–55). Although most physicists of the day regarded the lack of mathematical content to be a weakness of Faraday's work, the nontechnical character of Faraday's experimental reports struck Maxwell as an opportunity for assimilating Faraday's experimental results in an embryonic state. So attractive was this feature to Maxwell that he made a conscious decision to steer clear of any work on electricity by scientists who approached the subject principally through a mathematical analysis of centers of forces acting at a distance. This aversion to electromagnetic interaction of bodies acting at a distance was a consistent theme in Maxwell's work. His close friend, Peter Tait (1831–1901), in a review of Maxwell's *Treatise on Electricity and Magnetism* (1873) published in *Nature*, declared that the main objective of the work was to "upset completely the notion of *action at a distance*" (Tait 1873, 478).

Another factor that was critical in Maxwell's decision to focus on the work of Faraday was his conviction that the science of electricity needed to be grounded in the phenomena. This commitment to an empiricist methodology would later be reflected in the appointment of Maxwell at the age of 42 as first professor of natural and experimental philosophy and first director of the Cavendish Laboratory, when it was established in 1874.

Maxwell quickly integrated many of the central conceptual elements of Faraday's work into his own developing understanding of electromagnetism. One idea that he adopted straight away was that of *lines of force*, a concept, as we have seen, that was central to Faraday's later thought. Just as iron filings arrange themselves around a magnet, Maxwell held with Faraday, lines arrange themselves around a charge in a way that indicates the *direction* in which a point charge would move if it were to be introduced at any location. This idea highlighted the role played by the space around charges, thereby introducing the idea of a potential field. By definition, a potential does not actually exist until it is manifested in experiment. The field around a magnet, for example, has a potential that is not realized until an electric charge is placed in that field. Only by interacting with something can the potential of the field be manifested; until this time, it has only the potential to do something.

The notion of a potential had been developed by Laplace in the theory of gravitation and subsequently applied to electrostatics by Poisson and George Green (1793–1841). This work was extended by Green, a self-taught mathematician and pioneer in the application of mathematics to physical problems. It was Green who gave the name *potential* to this function in his landmark 72-page paper, "An Essay on the Application of Mathematical Analysis to the Theories of Electricity and Magnetism" (1828), which was published privately at Green's expense, because he thought it would be presumptuous for a person like himself, with no formal education in mathematics, to submit the paper to a reputable journal.

In Green's hands, the mathematical conception of potential became a powerful tool that facilitated the establishment of numerous theorems in electrostatics and magnetism, one of which is known today as Green's theorem. It relates the properties of mathematical functions at the surfaces of a closed volume to other properties inside. In its usual form, the theorem involves two functions, but it readily simplifies to what is often called the divergence theorem or Gauss's theorem. Although neglected for many years, Green's paper inspired much of the mathematical work of William Thomson in electrostatics and magnetism.

The concept of the potential came to have a crucial role in the mathematical theories of the ether and the electromagnetic field because the partial differential equation for the potential provides an expression for a theory of continuous action in a medium. Thus, Maxwell later explained that he considered the concept of the potential to be appropriate to the representation of Faraday's idea of lines of force traversing space, providing the basis for his theory of the field, in which contiguous elements of the field transmit force and energy (Maxwell 1873, 1:xi).

The notion of a field of potential contrasted sharply with action at a distance. Implicit in the model of action at a distance was the presence of physical connections between a magnet and a piece of iron; like a row of dominoes tumbling down to let one end of the iron rod know about the presence of a magnet at the other end, it was these connections that allowed contact over a distance to be made. As we shall see, the issue for Maxwell's developing field theoretical model was how did the field propagate—in empty space or through intangible air in the room? One explanation was to invoke the existence of an ether that filled space so the action could propagate through it.

Another idea adopted by Maxwell, though this time from a series of papers published in 1842 by William Thomson, was an analogy between charge distribution and heat flow. Consider a point source of heat P embedded in a homogeneous conducting medium. Because the surface area of a sphere is $4\pi r^2$, the heat flux ϕ through a small area dS at a distance r from P is proportional to $1/r^2$ in analogy with Coulomb's electrostatic law. Accordingly, by substituting "source of heat" for "center of attraction," a problem in electricity could be transformed into one in the theory of heat. Maxwell was careful to hold that the two sets of phenomena—conduction of heat and charge distribution—are dissimilar and that the analogy is based only on the mathematical resemblance of some of their laws.

Through a combination of this analogy and Faraday's lines of force, Maxwell attempted his first discourse on electricity—his 1856 essay "On Faraday's

Lines of Force" (in Maxwell, 1890, 1:155-229). Part I of this paper was read to the Cambridge Philosophical Society on December 10, 1855, and Part II on February 11, 1856. In this paper, Maxwell made use of Thomson's 1842 analogy between electric charge and heat flow, but in place of heat assumed "an imaginary fluid" in which the pressure varied inversely as the distance from the source of the fluid. Assuming positive and negative charges as sources and sinks of the fluid, Maxwell argued that the fluid would flow from source to sink along precisely the same lines as Faraday's lines of force. Furthermore, the lines of force (or, rather, the space between lines) could be regarded as thin tubes of steadily flowing, continuous, incompressible fluid. In that case, an electric current of strength u is connected to the magnetic field $\mathbf{H}$ to which it gives rise to the partial differential equation

$$\text{curl } \mathbf{H} = 4\pi u,$$

where the operator curl is an auxiliary vector that serves to give a quantitative representation of the way in which the magnetic lines of force curl about electric lines of force, in every point of space.

The notion of the lines of force forming a tubular surface was taken directly from Faraday. Maxwell used it to construct what he called a "geometrical model," which defined the motion of the fluid by dividing the space it occupied into tubes. The tubes were mere surfaces directing the motion of the fluid, which filled all space, and the forces were represented by the motion of the fluid. This model, Maxwell insisted, was not to be taken to be a physical representation of the lines of force; thus the fluid was "not even a hypothetical fluid" but merely a collection of imaginary properties for the expression of mathematical theorems (Maxwell 1890, 1:160). In Maxwell's view, this way of reasoning about phenomena, with its reliance on geometry and physical analogy, is to be contrasted with the expression of known results in terms of mathematical formula and the framing of physical hypothesis. Reasoning in this manner, Maxwell contended, was conducive to the production of "physical ideas without adopting a physical theory."

In the second part of his first foray in electromagnetism, Maxwell discusses how Faraday's electrotonic state might be represented by mathematical symbols. Electrotonic intensity is identified with the vector a, which Thomson had defined in 1847 in terms of the magnetic induction $\mathbf{B}$ (curl a = $\mathbf{B}$) and with the vector potential that others had employed in calculation of induced currents. In so doing, Maxwell, like Faraday in his later thought, assumes the primacy of lines of force. Although he did not discuss the nature of the physical state to which the lines of force corresponded, the lines of force were taken to represent a real physical state, and so were not understood to be mere theoretical entities.

This paper signified an important step: It brought Faraday's physical, geometrical conceptions under the control of powerful analytical mathematics.

One idea that Maxwell resisted in Faraday, however, was the identification of matter itself with the lines of force.

THE THEORY OF MOLECULAR VORTICES

The use of mechanical models and analogies was also a fundamental part of Maxwell's second foray into electromagnetism—his paper "On Physical Lines of Force" (1861–62, reprinted in Maxwell, 1890, vol 1, pp. 451–513). As the title suggests, with its allusion to Faraday's paper "On the Physical Character of the Lines of Magnetic Force" (1852; in *Experimental Researches in Electricity*, 1855, vol. 3, pp. 407-437), it was Maxwell's attempt to construct a mechanical model that would account for the strain, or tension, exerted on electrified bodies by means of the action between the contiguous parts of the medium in the space surrounding these bodies, rather than by direct action across the distance that separates them. This model, in turn, would permit a mathematical analysis from which certain deductions could be made. These deductions, finally, could be checked against phenomena.

The challenge, then, was to devise a mechanical model for the properties of the medium. Maxwell found the inspiration for his model in Thomson's explanation of the Faraday magneto-optic rotation: "Thomson has pointed out [in 1856] that the cause of the magnetic action on light must be a real rotation going on in the magnetic field" (Maxwell 1890, 1:505). In Faraday's experiments, the plane of polarization of a linearly polarized light wave passing through a piece of heavy glass in a strong magnetic field and in the direction of the field lines had been shown to rotate by an amount proportional both to the strength of the field and the distance traveled through the glass. This result had encouraged the consideration of magnetism as some sort of rotational state.

Thomson, in particular, insisted that the rotation could only be explained by the existence of some rotational motions induced in the medium by the magnetic field, centering on the magnetic lines of force as axes and with their direction determined by that of the field. Analyzing the motion of circularly polarized waves through an elastic solid medium, Thomson concluded that the observed effects could only be produced given the presence of molecular vortices set up and aligned by the magnetic force lines. No other explanation, he concluded, was conceivable.

Just as Maxwell's first paper can be viewed as a development of Thomson's heat analogy, this new investigation can also be viewed as a development of Thomson's molecular vortex theory. As before, however, Maxwell's second foray into electromagnetism was more than just an extension of another person's work.

Maxwell's model had two central features:

First, Maxwell conceives an electric current as represented by a transfer in a definite direction of these movable particles between neighboring vortices. In figure 9.1, the large spaces represent vortices and the small circles represent the

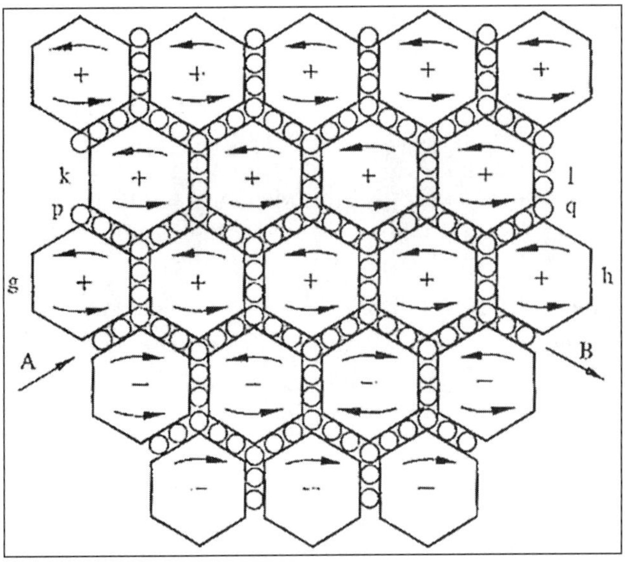

Figure 9.1: Array of vortices (end on). From Maxwell, "On Physical Lines of Force" (1856).

layers of particles between them, which, by hypothesis, represent electricity. If a current flows in the direction *AB*, the row of vortices *gh* will be set in motion in a counter clock-wise or positive direction. Supposing the row of vortices *kl* to be still at rest, then the layer of particles between these rows will be acted on by the row *gh* from beneath and will be at rest above. If free to move, these particles will therefore rotate in the clockwise or negative direction, and will move from left to right. This movement will be in a direction opposite to that of the original current, and from it will so form an induced current.

Maxwell therefore conceived the phenomena of induced currents to be due to the communication of the rotational velocity of the vortices from one part of the field to the other. Later, from the facts that the lines of force pass from the north pole of one magnet to the south pole of another, and that these poles attract one another, it was clear that the lines of force were lines of tension.

Second, the mutual lateral repulsion of the lines of force was reconciled with tension along the lines of force by the postulation of centrifugal force of vortices or eddies in the medium. These vortices have their axes in directions parallel to the lines of force, which creates an inequality of pressures that is consistent with the dipolar character of the lines of force. "Every vortex is essentially dipolar, the two extremities of its axis being distinguished by the direction of its revolution as observed from those points" (Maxwell 1890, 1:455).

One difficulty raised by this model was that, since all the vortices were assumed to rotate in the same sense round the lines of force, the contiguous parts of vortices round adjacent lines of force must be moving in opposite directions. The only way that he could conceive such a combination of motions was through something corresponding to what in engineering were termed *idle wheels* lying between the vortices. In figure 9.1, the hexagonal parts represent vortices, also called cells, and the circles represent the idle wheels.

From the properties of this model, Maxwell extracted a daring consequence: "The velocity of transverse undulations in our hypothetical medium ... agrees so exactly with the velocity of light ... that we can scarcely avoid the inference that light consists in the transverse undulations of the same medium which is the cause of electric and magnetic phenomena."

CLASSICAL FIELD THEORY

Maxwell's next contribution to the science of electromagnetism was his "A Dynamical Theory of the Electromagnetic Field" (in Maxwell 1890, 1:526–597). In this paper, published in 1864, Maxwell advanced a new theoretical framework on the basis of a few general dynamical principles and experiments, from which the propagation of electromagnetic waves through space followed without any special assumptions about molecular vortices or the forces between electric particles. Indeed, Maxwell formulated an electromagnetic theory that made no reference whatsoever to the microstructure of matter—the sole physical assumption in this framework was the existence of an ethereal medium:

> It appears, therefore, that ... there is an aetherial medium pervading all bodies, and modified only in degree by their presence; that the parts of this medium are capable of being set in motion by electric currents and magnets; that this motion is communicated from one part of the medium to another by forces arising from the connection of those parts. (Maxwell 1890, 1:556)

With this and the mathematical results of his earlier papers, Maxwell constructed a theory of the electromagnetic field, complete with a statement of the guiding principle of classical field theory: The energy of a physical system is not to be found in the material particles of which it is composed but in the ethereal medium surrounding these particles. This principle eliminated the need for "an unknown quality called potential, or the power of producing certain effects at a distance" (Maxwell 1890, 1:564). The field itself can be described "without [such] hypothesis as magnetic polarization and electric polarization, or, according to a very probable hypothesis, as the motion and the strain of one and the same medium" (Maxwell 1890, 1:564). Electromagnetic processes are transmitted by the separate and independent action of each charge (or magnetized body) on the surrounding space, rather than by direct action at a distance.

In the wave theory of light, it was necessary to posit a medium, or ether, and to endow it with certain properties in order to interpret the phenomena of refraction, polarization, and so on. Maxwell pointed out that it is "unphilosophical" to fill space with another kind of ether every time a new phenomenon had to be explained. It seemed reasonable to him that the same medium required for the wave theory of light also served for the explanation of electromagnetic effects. Assuming that an electromagnetic disturbance is propagated through the field by means of a plane wave (a wave in which the electric intensity is the same at any instant over the whole plane), he obtained a simple expression for the velocity of such a wave.

It seemed clear to Maxwell that the ether that contained the energy of electromagnetism was the same ether that transmitted the undulations of light. The only question was whether there was a close connection between electromagnetic and electrical phenomena. Maxwell's equations produced the result that the rate of

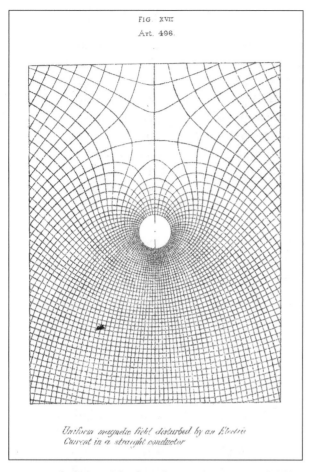

Figure 9.2: Drawing of the disturbance of a magnetic field by an electric current in a straight conductor. From Maxwell, *Treatise on Electricity and Magnetism* (1873).

propagation of the electromagnetic field through space was almost exactly equal to that of the velocity of light. The connection was clear: Light (including radiant heat, and other radiations if any) is an electromagnetic disturbance in the form of waves propagated through the electromagnetic field.

Maxwell's electromagnetic theory has assumed such a prominent place in the annals of physics that it is easy to overlook the original features of Faraday's field theory that Maxwell modified. The most fundamental modification was Maxwell's treatment of matter and the field as basically different entities. In Newton's theory, each particle exerts force at a distance on other particles. Because the action of one body on another is instantaneous, there is little reason to regard the field as more than a mathematical device for calculating force; Newtonian forces were thought to be caused directly by the distant bodies, not the field. In Maxwell's theory, one body does not act directly on another, but acts progressively through the field. However, there is a division between matter and force, just as in Newton's theory. The electric force does not, strictly speaking, consist of fields of force, but of forces per unit charge, and the field is assumed to exist at the same point as the charge it acts on.

Although Maxwell does assert in his 1861–62 paper that one is free to choose either magnetic and electric polarization or motion and strain of a medium, late in his life he made his convictions clear: "Whatever difficulties we have in forming a consistent idea of the constitution of the ether, there can be no doubt that the interplanetary and interstellar spaces are not empty, but are occupied by a material substance or body, which is certainly the largest, and probably the most uniform body of which we have any knowledge" (1890, 2:674).

One important consequence followed from Maxwell's calculations: The optical spectrum should be only a small region in a much larger range of wavelengths. There ought to be electromagnetic radiation beyond the optical range that could be created and detected by purely electromagnetic means. The detection of such radiation was left to others. We will look more closely at the study of electromagnetic waves in the next chapter.

ARTICULATING MAXWELL'S ELECTROMAGNETIC THEORY

Some historians have suggested, with some justice, that Maxwell's equations did for the field theoretic view of the universe what Newton's laws of motion had done for the mechanist view of the universe (e.g., Williams 1966, 136). Maxwell's equations summarized the fundamental relations between electricity and magnetism and electromagnetic radiation in the most elegant fashion imaginable. These equations expressed, respectively, how electric charges produce electric fields (Gauss's law); the experimental absence of magnetic charges; how currents produce magnetic fields (Ampere's law); and how changing magnetic fields produce electric fields (Faraday's law of induction). Nothing was omitted. What's more, as I have mentioned in passing, they hinted at the existence of electric waves propagated through dielectric media, based largely on Faraday's lines of force.

In the long term, Maxwell's equations were shown to be approximations that held for low velocities and short distances; if they were to apply with complete generality, they required the modifications of Einstein's relativistic physics. Still, Maxwell's equations have survived all the changes in knowledge occasioned in the twentieth century by relativity and the quantum theory. These equations are as valid today as they were when they were first stated nearly 150 years ago.

In retrospect, it is somewhat surprising to discover that many scientists were not persuaded when Maxwell's electromagnetic theory first appeared in the 1860s. As one might expect, Maxwell's electromagnetic theory received stiff opposition on the continent from proponents of action at a distance theories, which was abetted to some extent by the appearance of Helmholtz's law of conservation of energy in 1847. Helmholtz's theory still incorporated action at a distance between discrete electric charges but made provisions for a dielectric medium. However, many other scientists, who were not sympathetic to action-at-a-distance approaches to electromagnetic phenomena, were reluctant to throw their support behind Maxwell's theory. The failure of Maxwell's theory to accommodate phenomena, such as the reflection and refraction of light, may have played a minor role in the resistance of the scientific community to Maxwell's electromagnetic theory, but the central factor, if there was one, was the difficulty of the theory itself. The central concepts were difficult to grasp. Even Heinrich Hertz (1857–1894), who we will discuss at length in the next chapter, portrayed Maxwell's work as enormously difficult:

> Many a man has thrown himself with zeal into the study of Maxwell's work, and, even when he has not stumbled upon unwonted mathematical difficulties,

has nevertheless been compelled to abandon the hope of forming for himself an altogether consistent conception of Maxwell's ideas. I have fared no better myself. Notwithstanding the greatest admiration for Maxwell's mathematical conceptions, I have not always felt quite certain of having grasped the physical significance of his statements (Hertz 1893, 1).

A second related factor, mentioned by Hertz in this passage, in its becoming known and accepted by scientists of the day was the mathematical complexity of Maxwell's work. In his 1861–62 paper, Maxwell developed mathematical expressions by which the various manifestations of the energy of this field could be treated with quantitative exactness—some 20 equations in all with the same number of variables. These equations were pared down in his "Notes on the Electromagnetic Theory of Light" (1868; in Maxwell 1890, 2:125–43) and written in the integral form, based on four postulates derived from electrical experiments. In Maxwell's *Treatise* (1873), the number of equations describing the electromagnetic field was reduced to eight. Still, it required several other individuals—Oliver Lodge (1851–1940), G. F. FitzGerald (1851–1901), Oliver Heaviside (1850–1925), and Joseph Larmor (1857–1942)—working for some years after Maxwell's death in 1879, before this innovative work was interpreted, simplified, and distilled into the set of rules we now know as Maxwell's equations. It was this version of his theory that possesses the virtues of compactness and analytical symmetry, which has been passed into the textbooks as Maxwell's electromagnetic theory.

The real appeal of Maxwell's theory was to be found, not so much in his derivation of electromagnetic properties as in the new foundation he provided for the theory of light. Fresnel's wave theory of light had enjoyed great success, but all attempts to position this theory upon a model of the luminiferous ether had engendered difficulties. In particular, extant ethereal models did not discriminate between transverse and longitudinal vibrations, and so were unable to explain why experiment had revealed only the former type. These models also failed to reproduce Fresnel's result for the double refraction and reflection of polarized light in the interface of two optical media. Maxwell's electromagnetic theory, in contrast, predicted that only transverse vibrations could be propagated through the electromagnetic medium and led to a simple derivation of Fresnel's results.

Maxwell did not publish this derivation, but he pointed the way toward it. In his doctoral dissertation (1875), the Dutch physicist Hendrik Antoon Lorentz (1853–1928) derived Fresnel's results from Maxwell's theory and showed explicitly how this theory accounted for the absence of longitudinal light waves. Later in 1878, Lorentz extended his analysis into dispersion theory, where once again the ether models failed but Maxwell's electromagnetic theory succeeded. Finally, in the 1880s, Lord Raleigh (1881) and Josiah Willard Gibbs (1888) showed that Maxwell's system of optical equations was the only one to give a self-consistent set of formulae for reflection, refraction, and dispersion in agreement with experiment.

These researches furnished evidence for Maxwell's electromagnetic theory of light but not of a conclusive nature. The purely electrical consequences of Maxwell's theory were in agreement with all known electrical observations. The equations of the field accounted for the electromagnetic forces observed in various experiments, and from them the laws of electromagnetic induction could be deduced in a straightforward manner. However, there were no direct measurements of the velocity of electric waves. The velocity that they must travel, if Maxwell's theory were true, had been determined from observations. The velocity of light was known from the experiments of Armand Fizeau (1819–1896) and Jean Foucault (1819–1868) to have approximately the same value. Indeed, it was this near coincidence that led Maxwell to write in 1865 "the agreement of the results seems to show that light and magnetism are affections of the same substance, and that light is an electro-magnetic disturbance propagated through the field according to electro-magnetic waves" (1865, in Maxwell 1890, 1:565).

As we shall see in the next chapter, it was the research of Heinrich Hertz that was the decisive factor in the triumph of Maxwell's theory. Maxwell told the scientific community what the properties of electromagnetic waves in air must be. But it was Hertz who in 1887, measured these properties, and these measurements completely verified Maxwell's views.

POSTSCRIPT: THE DISPLACEMENT CURRENT

Maxwell struggled with the paradox of the capacitor, where charge entered one plate and then flowed out of the other plate apparently without traversing the space between the plates. It appeared as though electric charge was being destroyed on the upper plate and being re-created when it reappeared on the lower plate. To formulate electrodynamics in such a way that it did not rely on action at a distance, Maxwell postulated a new type of current, called "the displacement current," in his third paper on electromagnetism.

The displacement current is represented in Maxwell's first equation as the complicated-looking last term in Ampère's law. Ampère's original law, converted to differential form in terms of the magnetic field **B**, was

$$\nabla \times \mathbf{B} = \mathbf{j}$$

Where $\nabla \times$ is the curl operator, **B** is the magnetic field strength, and **j** is the current density, understood to represent the flow of electric charge. His claim was that the current density **j** at a given location does not actually represent the total current flow at that location (even though that is its definition).

According to Maxwell, a dielectric medium can be considered to consist of couples of positive and negative charges, and an electric field **E** pulls these charges in opposite directions, stretching the links between them until they achieve some kind of equilibrium. If the strength of the field is increased, the charges are pulled further apart, so during periods when the electric field is changing, there is movement of the electric charge of the dielectric medium.

This movement of charge is the displacement current, proportional to $\nabla \mathbf{E}/\nabla t$, which he adds to Ampère's original formula to give

$$\nabla \times \mathbf{B} = \mathbf{j} + \frac{\partial \mathbf{E}}{\partial t}$$

The introduction of the displacement current is perhaps the single most remarkable feature of Maxwell's electromagnetic theory, principally because the introduction of the term $\partial \mathbf{E}/\partial t$ in Ampere's equation leads to transverse electromagnetic waves propagating in a vacuum at the speed of light. With the addition of the displacement current, the equation states that a changing electric field (right-hand side of equation above) creates a magnetic field (left-hand side of equation). Faraday's law states that a changing magnetic field creates an electric field. With Maxwell's modification, the two laws are almost exact mirror images of each other, but did this symmetry mean anything? There was now reason to suppose that electric fields and magnetic fields, once they are set up correctly, just keep working back and forth, each one changing and creating the other in turn, all by themselves with absolutely nothing else to help them along.

The historical literature tends to present this symmetry as a *consequence* of the added term, and indeed Maxwell himself presents it in this way. However, it would not be unkind to point out that the existence of electromagnetic waves was also a possible motivation for Maxwell's modification of Ampère's law. One would assume that Maxwell must have been alert to the fact that, if Ampere's law contained a term of the form $\partial \mathbf{E}/\partial t$, the standard wave equation would follow from the known system of electromagnetic equations. The objective of explaining the wave properties of light was certainly in the air at that time. Already Faraday had speculated that the electromagnetic ether and the luminiferous ether might well turn out to be the same thing, suggesting that light actually is a propagating electromagnetic disturbance. Also, Wilhelm Eduard Weber (1804-1891) had shown that a speed on the order of the speed of light is given by a simple combination of electromagnetic constants. Other people had pursued the same idea.

10

ELECTROMAGNETIC WAVES

INTRODUCTION

In considering the implications of his equations, Maxwell found that a changing electric field had to induce a changing magnetic field, which, in turn, had to induce a changing electric field, and so on. The field progressed outward in this manner in all directions. The result was a radiation possessing the properties of a waveform. Maxwell had predicted, in short, the existence of electromagnetic waves—many kinds of electromagnetic waves, and not just light waves—with frequencies equal to that in which the electromagnetic field waxed and waned. In particular, Maxwell's theory predicted the existence of waves of longer wavelength, traveling at the same speed as light, which scientists might be able to generate artificially using electric currents in wires. Just such waves were generated with an induction coil and studied by Hertz in the 1880s. Hertz was an ingenious experimenter, and he was compelled to develop entirely new techniques to detect the electromagnetic waves predicted by Maxwell. These new waves are now called radio waves.

ELECTRIC WAVES

Hertz was first attracted to Maxwell's equations in 1883 when Hermann Helmholtz, his professor at the University of Berlin, suggested that Hertz try for a prize offered by the Berlin Academy of Science for work in electromagnetism. The prize was offered to the student who supplied the best answer to the question: Did electricity move with inertia? In other words, when a circuit was broken, would electricity continue to move? How long did it take to react when a current was started or stopped?

To measure currents in his experiments, Hertz employed a homemade galvanometer, which was fixed to a wall. He built his own batteries and sent the current through wires that led through the wall, effectively insulating him from the poisonous fumes. He was informed in August 1879 that he won the prize by showing that electricity had no inertia. The award marked him and his research in the eyes of his colleagues as a powerful, new force in the profession. In 1883, Hertz left Berlin and eventually settled in Karlsruhe, where he accepted a position as professor of experimental physics at the Technische Hochschule in 1885. This university was the site of his celebrated experiments on electric waves.

Hertz had already published a number of theoretical papers, attempting to find proofs for Maxwell's theory and to connect it to the electrodynamic theories of Wilhelm Weber (1804–1891) and Carl Neumann (1832–1925). In these theoretical studies, the existence of electromagnetic waves loomed large. It was also clear that the frequency of electromagnetic waves was of paramount importance in any experimental investigation, and with this in mind, Hertz set out to find a way of obtaining very high-frequency oscillations.

In 1886, seven years after Maxwell's death, Hertz tried unsuccessfully to determine the influence of dielectrics on the induction of sparks between a primary circuit and a secondary detection circuit. The following year, he turned his attention to the induction of sparks through air but across a relatively large distance, and to the measurement of the velocity of propagation of the electric waves concerned. If this velocity were found to be finite, it would provide evidence for the existence of a medium, such as that hypothesized by Maxwell. Helmholtz had attempted to measure this velocity without success.

Hertz began to experiment with sparks in a short magnetic loop attached to an induction coil. By connecting its primary winding to a source of electricity, he found that it was possible to induce far greater voltage in the secondary winding. If the secondary winding was left open and allowed to terminate at two metal balls separated by a small distance in air, impressive sparks could be generated.

In setting up his apparatus, Hertz employed induction coils that he found in an old closet at the university. These coils were commonplace items in late-nineteenth-century laboratories, but Hertz put them to a different use. He connected two metal rods with spheres on their ends to the two sides of the spark gap, so that he could vary the electrical characteristics of the circuit (fig. 10.1). In effect, he had constructed a crude radio antenna.

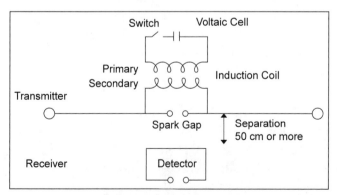

Figure 10.1: Diagram of Hertz's apparatus, which he used to explore the so-called electric waves that were generated when the induction coil at the left sent a spark between the two small spheres at the center. Artist: Jeff Dixon.

On November 1, 1886, his apparatus, or oscillator, caused sparks to jump across the gap. What caught his attention was the presence of "side sparks" that were somehow produced by the large spark in his apparatus. These side sparks were very small—perhaps only a hundredth of a millimeter long—and they could have easily been overlooked. Seeing the sparks at all required keen eyesight, pitch black, and immense good fortune.

Hertz set himself the task of unraveling the mystery of the side sparks. He found that he could produce sparks in a secondary circuit at will, provided that it was placed in close proximity to a primary circuit that was sparking. In so doing, he was able to bring the phenomenon of the side sparks under control. It was this control that afforded him the chance to ascertain their characteristics and isolate the factors that were at play in their production.

To this end, he constructed a detector that consisted of a small metal loop terminating in a spark gap, the spacing of which could be minutely adjusted with a micrometer. With this simple but crude detector, he determined where the sparks were the longest and how their length changed with position. From this measurement, he knew how much energy was present at any given location to produce the side spark. When he graphed the length of spark at different positions, it produced a wave pattern running from zero to a maximum and then back to zero again (fig. 10.2). Hertz concluded that an invisible wave, or "electric wave," was present in his apparatus. The detector mapped out the shape of the wave, from zero to high and back to low and zero again, as the detector was moved away from the primary spark.

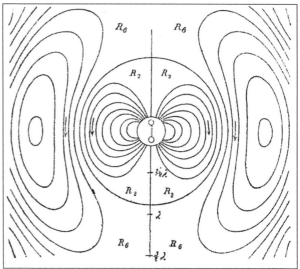

Figure 10.2: Hertz's map of electromagnetic waves. The two small circles in the center of the illustration represent two metal balls, separated by an air gap. From Hertz, *Untersuchungen uber der Ausbreitung der elektrischen Kraft* (1892).

The distance between the locations where the spark was nonexistent gave the length of the wave. From the spark discharge, Hertz ascertained the frequency of the wave. He then combined frequency and wavelength in a simple formula and determined the velocity of the wave. He now knew how fast the wave was moving—that it was moving at the speed of light but was not light. Hertz soon realized that he had discovered a new form of electromagnetic radiation that had some of the properties of light but a much longer wavelength. This discovery was the confirmation of Maxwell's prediction.

Hertz began to explore the similarities between light and his electric waves, showing that they could be refracted (bent) in passing through materials such as pitch, diffracted (broken apart) in passing through narrow holes in metal screens, and polarized (acquire a new orientation) by passing them through a grid of parallel wires. He even learned how to focus his electric waves using large concave mirrors, the forerunner of the metal dishes that are now a staple in satellite communication. Hertz concluded that the two types of waves were identical.

Oliver Lodge (1851–1940), who also worked hard to verify Maxwell's theory, attempted to generate waves along an open transmission line. In 1887–88, he showed that a strong spark could be produced at the far end of a transmission line only when the line was resonance. Lodge obtained resonances with wires 95 feet long, corresponding to a resonant frequency of about 5 MHz. He expected that he would be able to report his results, which supported Maxwell's theory, at the September 1888 meeting of the British Association. However, Hertz published his result first. Hertz acknowledged that "there can scarcely be any doubt that if I had not anticipated him, he [Lodge] would also have succeeded in observing waves in air, and thus also in proving the propagation in time of electric force" (1893, 3).

The production of electric waves by electric means was recognized as an important achievement immediately. One factor was the ease with which the experiment could be reproduced. The required equipment was readily available. Hertz even provided instructions on how the scientist could overcome any difficulty pursuant to the adjustment of the primary spark. By 1889, Hertz's result had been reproduced many times. These new invisible electromagnetic waves were originally called Hertzian waves, and they conclusively confirmed Maxwell's prediction of the existence of electromagnetic radiation, both in the form of light waves and of radio waves. The waves were later renamed radio waves, when the Italian Guglielmo Marchese Marconi (1874–1937) adapted Hertz's technology for wireless communication in the 1890s.

SIMPLIFYING MAXWELL'S EQUATIONS

Hertz himself quickly converted to Maxwell's theory but not to all of it. When he started working on Maxwell's electromagnetic theory, he had accepted Maxwell's equations but had rejected his hypothesis of an ethereal medium. After he generated electromagnetic waves, Hertz accepted the medium but had reservations about the equations. Moreover, while he accepted the medium as a fact of the matter, Hertz was not happy about treating it as a fundamental hypothesis.

Over the next few years, Hertz reformulated Maxwell's theory in simplified, axiomatic form. He eliminated Maxwell's vector potential **A**, which he regarded as a relic from action at a distance theories. In the case of gravitation, for example, the attraction at any point due to a given distribution of mass can be expressed as the force that would act on a body if it were to be placed there. The gravitational potential describes a potential force that would be exerted if a small mass were to be introduced at any point. This potential need not describe a material property of the gravitational field. If our frame of reference is Maxwell's ether, however, this expression describes an actual property of material at every point in the field. Hertz concluded that the fundamental equations should describe relations between physical magnitudes that are actually observed, and not between magnitudes that are only useful for calculation.

Also eliminated was the distinction between electric displacement and electric force in the free ether, which again only made sense on the assumption of action at a distance. He then reduced Maxwell's full system of equations as presented in "A Dynamical Theory of the Electromagnetic Field" (1865) and the *Treatise* (1873) to a symmetric system of equations. In the introduction to *Electric Waves*, Hertz maintained that "Maxwell's theory is Maxwell's system of equations" (1893). What he actually meant by this cryptic remark was his own simplified version of Maxwell's system of equations. Hertz took these equations as fundamental and left their interpretation to others. Maxwell's full theory was clearly richer than Hertz's simplistic formulation, but it was flawed in its treatment of charge; Hertz's way to remove this flaw was to start with the equations and work up a self-consistent representation of them.

THE ETHER AND THE ELECTRON

Maxwell's theory of electromagnetism was taught to undergraduates as early as 1871, when Maxwell accepted an appointment as professor of experimental physics at the University of Cambridge. It was only a decade later, however, that his scientific work became the stimulus for the ongoing research of others. For example, one of the first publications of J. J. Thomson (1856–1940)—an outstanding Cambridge graduate who would succeed Maxwell as director of the Cavendish Laboratory—was a study of the electromagnetic effects produced by the steady motion of a charged conductor according to the principles advanced by Maxwell in his *Treatise* (1873).

The guiding principle of Maxwell's theory was that all electromagnetic phenomena are to be attributed to mechanical processes taking place in the ether. The electromagnetic field, in Maxwell's view, is not an independent dynamical reality but one of the mechanical states of the ether. A current in a wire, for example, was a spontaneous breaking down of the electric tension, or displacement, in the ether in the vicinity of the wire. Just how the presence of the wire caused the electrostatic energy in the ether to be converted into heat was left unexplained by Maxwell.

Hertz's discovery of electromagnetic waves in 1888 is sometimes portrayed by historians as having confirmed the theory of the electromagnetic field. At the time, however, scientists believed that Hertz's discovery confirmed the existence of the ether as a mechanical substance; in other words, they interpreted

Hertz's experiment as furnishing experimental proof for the existence of an ether. It is for this reason that followers of Maxwell in the late-nineteenth century dedicated their energies to the construction of mechanical models of the ether.

The guiding principle of Maxwell's theory was challenged by another outstanding graduate of Cambridge—Joseph Larmor—who would go on to become Lucasian professor of mathematics at Cambridge in 1903. Larmor's interest in electromagnetic theory was stirred by a paper published by the Irish physicist George FitzGerald, one of a handful of scientists outside of Cambridge who had discovered Maxwell's *Treatise* and were engaged in the attempt to articulate his theory more completely. During the period from 1894 to 1897, Larmor's paper was published in three installments (with assorted appendixes) with the title "A Dynamical Theory of the Electric and Luminiferous Medium" in the *Philosophical Transactions of the Royal Society*. These three papers presented Larmor's theory of the electron.

FitzGerald acted as referee for this paper. He requested and received permission from the publisher to contact the author directly, encouraging Larmor to introduce the concept of the "electron" into his theory. Larmor acquiesced, which, in turn, exerted a profound influence on Larmor's understanding of the relationship between the electromagnetic ether and gross matter. If matter itself was composed of positive and negative electrons, in keeping with FitzGerald's suggestion, then virtually every problem, both in electrodynamics and matter theory, would be transformed into a problem in the electrodynamics of moving bodies. Such previous well-known effects as the electric polarization and magnetization of matter, which had been attributed to changes to the dynamical properties of the ether, could now be explained in terms of the material structure of matter.

The publication of Larmor's papers played a significant role in the fortunes of Maxwell's electrodynamic theory. Prior to the appearance of Larmor's first paper in 1894, physicists on both sides of the Atlantic were enthusiastic in their endorsement of Maxwell's theory. After the publication of Larmor's third paper in 1897, few physicists (a notable exception being Oliver Heaviside) openly supported Maxwell's electromagnetic theory. Indeed, it was during this critical three-year period from 1894 to 1897 that the guiding assumptions of Maxwell's theory gave way to the novel proposition, courtesy of Larmor in consultation with FitzGerald, that the only source of charge is a particle, that the flow of such particles uniquely constitutes the current of conduction, and that the ether must be strictly separated from ponderable matter.

The currency that Maxwell placed on his belief that all electromagnetic phenomena are to be attributed to mechanical processes taking place in the ether was challenged on a second front by the Dutch physicist Hendrik Antoon Lorentz, who advanced a startling new account of the relation between the velocity of light and matter in two papers of 1892 and 1895. Concerns about this relationship had arisen in the context of the explanation of stellar aberration according to the wave theory of light—the observation that stars do not always seem to be where they belong. If, however, light waves traveled through a medium, and the medium itself moved, then starlight reaching Earth might be affected by the ether.

Two possibilities arose. The first possibility, which was proposed by Fresnel, was that the Earth and other bodies passed unhindered through the stationary luminiferous ether, creating an "ether wind" across the Earth's surface. Fresnel's theory furnished a satisfactory explanation of aberration, but it clashed with the generally accepted view that to convey transverse rather than longitudinal waves, the ether would have to be a certain kind of elastic solid. The second possibility, proposed in 1849 by the Irish mathematician G. G. Stokes (1819–1903), was that there must be friction between the ether and the Earth, and that the ether must therefore be dragged along by the moving Earth, at least close to its surface.

Maxwell suggested in 1878 how one might measure the movement of the Earth relative to the ether by observing and comparing the velocity of light in different directions. He claimed, moreover, that no method would be able to detect the subtle difference involved; that is, it was not possible to document the movement of the Earth through space. Challenged by Maxwell's comments, Albert Michelson (1881) attempted to detect the relative motion of Earth and ether using an optical interferometer. Basically, the speed of light was measured in two perpendicular directions—parallel and perpendicular to the Earth's motion through space. If the ether existed, in one case the light would be traveling against an ether "headwind" and in the other it would be moving "crosswind." The two measured velocities would differ. Surprisingly, no difference in velocity was detected, in seeming vindication of Stokes's theory. Michelson repeated his experiment in 1887, this time with better equipment and the assistance of the chemist Edward Morley (1838–1923). Once again, the result did not point to any motion of the Earth through the ether.

Lorentz turned his attention to the subject of the ether drag in 1892 and to electromagnetic theory in general. Where Maxwell had attempted to formulate a pure field theory, the heart of Lorentz's theory was the distinction between matter and the ether, and the claim that a moving ponderable body cannot communicate its motion to the surrounding ether. The consequence was that no part of the ether could be in motion relative to any other part. Although it is immobile, the ether is not a fluid or a solid, endowed with such material qualities as density and elasticity, but is empty space itself with purely electromagnetic properties that are set out by Maxwell's equations. The ether, in Lorentz's view, is not a kind of mechanical substance but rather a dynamical medium for electromagnetic actions. In other words, the ether is an independently existing electromagnetic field.

This nonmechanical conception of the ether as an electromagnetic field raised a question that lies at the heart of Lorentz's electron theory; namely granted the evidence that there is an interaction between ordinary ponderable matter and the electromagnetic field, what possible attributes of matter make it capable of interacting with the electromagnetic field? Lorentz submitted, as answer to this question, that all ponderable bodies consist of minute particles that bear positive or negative charges, and that electric phenomena are occasioned by the displacement of these particles. Electricity, in Lorentz's view, consists of material particles with a definite charge and mass; that is, it consists of electrons.

In conformity with his basic assumptions, Lorentz demonstrated the power of his new approach by deriving the Fresnel drag coefficient. In a medium of refractive index, v, in which all charged particles share a common translational velocity, p, Lorentz showed that the velocity of propagation is in agreement with Fresnel's hypothesis (see below).

Lorentz's equation: $\frac{u}{v}+(1-\frac{1}{v^2})$, Fresnel's coefficient: $(1-\frac{1}{v^2})$

For Fresnel, the coefficient stood for a genuine dragging effect of the ether on a moving body, a view which led to consequences that were patently absurd. It implied, for example, that one and the same ether would be dragged by different ratios for rays of different wavelengths. Because the coefficient in his theory does not conform to a mechanical dragging effect, but is merely an interference effect between the incident light and that produced by the consequent vibrations of electrons in the optical body, Lorentz managed to sidestep this absurdity.

At the time of publication of his 1892 paper, Lorentz was aware of the Michelson-Morley experiment. Although this null result was clearly unfavorable to the hypothesis of a stationary ether, Lorentz retained the ether simply because of the successes of his new approach, especially its treatment of Fresnel's drag coefficient. Rather than abandon the ether, he elected instead to seek a way of eliminating the tension between Fresnel's theory and the Michelson-Morley experiment.

In his second paper of 1895, Lorentz interpreted the null result of the Michelson-Morley experiment as being due to a contraction effect—that is, moving bodies contract in the direction of motion—which cannot be measured because the measuring rods shrink in the same proportion. If the molecular forces in the body behave in the same way with respect to the ether as do the electrical forces, a fairly reasonable hypothesis, then a body moving through the ether must contract, and this contraction would be such as to compensate for the optical Fresnel drag effect due to the same motion through the ether, giving the observed null result. This hypothesis that objects and observers traveling with respect to a stationary ether undergo a physical shortening had been advanced a few years earlier by FitzGerald, but it was Lorentz who developed the idea mathematically into a much more robust description. As we shall see, this work, which became known as the FitzGerald-Lorentz contraction was a significant step toward Einstein's theory of special relativity (1905). Einstein's theory was premised on an identical contraction, but it abandoned the last plank of Maxwell's electromagnetic theory—the imponderable ether.

CONCLUSION

By 1895 or so, electron theory had emerged as a fruitful and promising area of physics. Physical theory was now dominated by two kinds of theoretical entities: the stationary ether and charged particles. As for the study of electricity and magnetism, electromagnetic phenomena were now understood in terms of the dynamical state of an ether devoid of mechanical properties.

As the expression "theoretical entity" suggests, the electron theory lacked experimental support. Many scientists suspected that the cathode rays that had been recently detected in vacuum tubes consisted of streams of charged particles, but this suggestion was sternly opposed by others who contended that this new type of ray was best explained by a wave theoretical model of some sort.

POSTSCRIPT: THE INVENTION OF WIRELESS TELEGRAPHY

Almost immediately after the voltaic pile was used to decompose water, scientists began to wonder whether voltaic electricity could be harnessed to convey signals at a distance. The enthusiasm for a method that would allow people to communicate with one another at a distance was increased by Oersted's discovery of electromagnetism in 1820. Ampère, for instance, pointed out the possibility of making an electric telegraph with needles surrounded by wires.

In 1836, Samuel Morse (1791–1872) constructed his first telegraph from an old picture frame and four years later took out a patent for "telegraph signs." He eventually persuaded Congress to construct the first telegraph line between Washington and Baltimore and, in 1844, transmitted his first message: "What hath God wrought." With his assistant, Alexander Bain (1810–1877), he invented the Morse code. A little later, Alexander Graham Bell (1847–1922) was able to transmit the first complete sentences over a telephone to his assistant: "Mr. Watson, come here, I want you." In 1884, the first telephone line between New York and Boston was opened to the public. By 1890, 50,000 Americans had installed telephones in their houses or offices.

Hertz didn't think his experiment had any practical applications, but others did, notably, Marconi, who became interested in Hertz's invisible rays to signal Morse code. He received little encouragement in Italy, and so set out for London, securing the interest of government officials from the war office, the admiralty, and the postal service. In 1897, he helped to form the Wireless Telegraph and Signal Co. Ltd., which in 1900 became Marconi's Wireless Telegraph Co. Ltd. He achieved the first international wireless transmission between England and France, arousing considerable attention from the press. By 1901, he was able to successfully transmit messages in Morse code across the Atlantic. His wireless communication involved no expensive wire or cable, and it was even faster than the telegraph. One of the beneficiaries of Marconi's work was the shipping business. Radio operators aboard ships sent messages to the shore for passengers and in emergencies could be the only hope of contacting rescue ships.

Radio transmission of Morse code was certainly useful, even life saving, but others began to wonder if it could be used to transmit other sounds, such as the human voice. Marconi believed waves were generated by creating a spark that caused a whiplash effect. A Canadian, Reginald Aubrey Fessenden (1866–1932), rejected this hypothesis, suggesting correctly that sound waves continuously rippled outward—in the way that water ripples outward when a rock is dropped into it. Additional experiments led him to suggest that, if the waves could be sent at a high frequency, it would be possible to hear only the "variations due to the human voice." On Christmas Eve, 1906, he transmitted the first music and voice program. It originated in Massachusetts and was received as far away as the state of Virginia.

Although many inventors thought of radio as a substitute for the telegraph or telephone, which transmit information from one point to another point, the fact was that anyone with a radio receiver could listen in on these "private" communications. Pretty soon, the lack of privacy in radio was turned into a benefit. Westinghouse, a company that manufactured radio receivers, decided to set up its own station in Pittsburgh, Pennsylvania, to broadcast information to everyone. Westinghouse was granted the first U,S, broadcasting license for its station, KDKA, in October 1920 and on November 2, 1920, KDKA held the first scheduled public broadcast.

As more and more people began listening in to radio broadcasts, inventors sought ways to design better receivers. Using the new technology of electron tubes, engineers introduced more sensitive receiver technology, such as the regenerative and superheterodyne, a subtle and elegant technique for improving reception and tuning at the same time. It is quite difficult to build an amplifier that will work at high frequencies, such as the ones radio uses, and also difficult to build a tuning filter than can select a narrow band of frequencies and yet be adjusted across a range of frequencies. The filter must tune in one station and reject all others, but then change to tune in other stations. It is comparatively easy, however, to build a tunable oscillator. If the oscillator signal is added to an incoming radio signal, a beat signal will result that has a frequency of the difference between the two. A fixed filter can be built to narrowly select this beat frequency and pass it on to a low-frequency amplifier. As the oscillator frequency is varied, different radio frequencies will be moved down to the beat frequency and so selected. In other words, a variable oscillator and a fixed, narrow filter can do the work of a variable narrow filter. This technique is used in practically every radio to this day.

In addition to being the first to transmit voice over radio, Fessenden developed amplitude modulation. AM, as we call it today, was a better way to broadcast voice and music than the technologies that came before, which were designed for Morse code. However, it was not without its problems, the principal one being that the receiver often hears a lot of noise along with the broadcast. The noise comes from sources in the atmosphere, such as lightning, which is why it is sometimes called *static*. American inventor Edwin H. Armstrong (1890-1954) came up with a different system—frequency modulation, or FM. By the 1930s Armstrong was able to improve the sound quality of radio transmissions by using FM. It would be many more years, however, before ordinary people began listening in. For the time being, AM still ruled the roost.

By the 1940s, radio broadcasting was a powerful communications tool and had reached its golden age. The programming on radio was a lot like today's television, with news, sports, dramas, comedy shows, and soap operas—not to mention commercials. Just as people today spend their evenings glued to the television, people gathered in their living rooms to listen to the radio.

There were almost no portable radios, except in cars. By the 1930s, they were the single most popular item of optional equipment in cars. After the 1947 invention of the transistor, however, radios shrank to the point where they could truly be taken anywhere. The transistor also made it possible to combine AM and FM radios (and later tape and compact disc [CD] players) into a single, small package.

Despite competition from television and the Internet, radio remains a major source of information and entertainment. Some radio talk-show hosts have tens of millions of listeners and for many individuals, radio remains their primary resource for music.

11

CHARGED PARTICLES OF MATTER

A FOCUS FOR RESEARCH

Science lecturers who traveled from town to town in the mid-nineteenth century delighted audiences by showing them the ancestor of the neon sign. The interior of a glass tube with wires embedded in opposite ends would glow in lively patterns when the air was pumped out of it and a high voltage was run across the wires. The lecturers had absolutely no idea what was going on in their glass tubes, but they were transfixed by this fluorescence.

This phenomenon of electrical excitation in a glass tube had been described by Hauksbee in 1709. Faraday himself devoted two series of his *Experimental Researches* (1839–55: numbers 1526 and 1552) to electric discharges in a vacuum. He noticed that the negative pole became covered with a continuous glow and that it was separated from the positive pole by what later became known as Faraday's dark space. Faraday was not able to pursue the matter further because the air pumps in use at the time were not up to the task.

The first good vacuum tubes were developed in 1855 by the German inventor Heinrich Geissler (1814–1879). As the air is pumped out, a stage is reached when a difference of potential of some 10,000 volts leads to a thick, furry-looking spark. Still further exhaustion to about a ten-thousandth of atmospheric pressure leads to a bright green glow, or fluorescence of the glass. Using a Geissler tube, as this pump was known, in 1858 the German physicist Julius Plücker (1801–1868) saw that where this light from the cathode reached the glass it produced a greenish glow. Many lines of research were pursued in the new laboratories that were created during the second half of the nineteenth century. However, research related to the green glow generated in evacuated tubes came to dominate the others and eventually culminated in an entire new theory of matter and a new understanding of current phenomena.

The German physicist Eugen Goldstein (1850–1930) set events in motion in 1871 when he suggested that this luminescence was caused by rays coming from the negative electrode and striking the glass. The negative electrode was commonly known as the cathode, and so Goldstein called these emanations cathode rays. During the course of intense experimentation, he found that the path of the rays could be altered if they were subjected to the influence of a magnet. The action of an electric field, however, seemed to have no effect on the rays. Because the waves were similar to light, Goldstein argued that a wave motion in the ether seemed to be the best explanation for these results. He also observed in 1886 that a cathode-ray tube produces, in addition to cathode rays, radiation that traveled in the opposite direction—away from the anode; these rays were called canal rays because of holes (canals) bored in the cathode; later these would be found to be ions that have had electrons stripped in producing cathode rays.

This view received much support from experiments carried out by Hertz in 1883, which indicated that cathode rays were not deflected by electrically charged metal plates. This result was taken to indicate (incorrectly) that cathode rays could not be charged particles. In another series of experiments, Hertz and his student Philipp Lenard (1862–1947) placed thin foils of gold and aluminum in the path of the rays and saw that the glass still glowed, as though the rays slipped through the foil in the manner that seemed consistent with a wave interpretation. The cathode rays emerged from these thin metal sheets in a disturbed state, like light passing through a turbid liquid, but it seemed inconceivable to Hertz that particles of matter should be able to cross the thinnest of metals. For this reason, Hertz held fast to the view that cathode rays must be a disturbance in the ether.

In 1870, Cromwell Varley (1828–1883) advanced a completely different interpretation based on his belief that cathode radiation was caused by the collisions of particles. Because it had been shown that the path of the rays was deflected in the presence of a magnet, these particles had to be considered carriers of an electrical charge (negative). If the rays really were made of charged particles, as Varley believed, one would expect that they should be deflected by the presence of an electrical field as well. Despite his best efforts, Varley was unable to orchestrate any such deflection.

William Crookes (1832–1919) designed a string-and-sealing-wax arrangement, known as a Crookes tube, to examine the behavior of cathode rays at various gas pressures, and to subject them to a magnetic field. In figure 11.1, the cathode is at the bottom of the glass tube. Directly above it is a mica screen with a central slit, through which a narrow beam of cathode rays can pass. In operation, the rays bounce off the walls of the tube and fall on the fan-shaped fluorescent screen, which is inclined at an angle to their path so that the waves produce a luminous line—by exactly the same principal that a picture is produced on a television screen. At the bottom of the stem of the tube is some caustic potash, which gives off moisture when heated and reabsorbs as it cools. By controlling the temperature of the tube, Crookes was able to increase or decrease the amount of

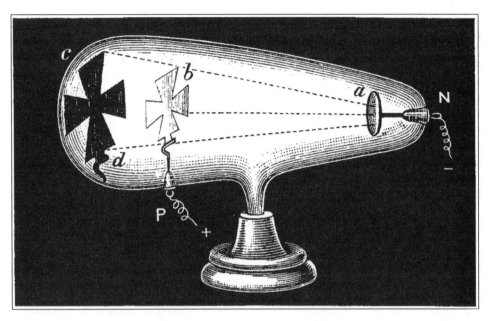

Figure 11.1: Crookes tube. From Crookes, "Radiant Matter" (1881).

the water vapor, and hence the vacuum, at will. In effect, by reducing the number of molecules present by lowering the pressure, cathode rays (electrons) can travel comparatively great distances without encountering atoms and being absorbed as they would be under ordinary atmospheric pressures.

Crookes established great facility in controlling the path of the rays. He found that when a bar magnet was brought near the tube, the beam of cathode rays was directed into a spiral, and that a horseshoe magnet bent the path into a curve. In this way, he gave experimental support to Varley's particulate theory of the cathode ray—the direction of the deflection was that which could be expected if the cathode rays consisted of a stream of negatively charged particles. Crookes also observed that the curved path produced by a magnetic field grew flatter as a better vacuum was achieved, demonstrating that the initial velocity of the particles increased as the vacuum became purer. In the glow of his tubes, Crookes believed that he had identified a world where matter may exist in "a fourth state" (as opposed to the previously known states, the solid, the liquid, and the gaseous), where the corpuscular theory of light may be true, and "where Matter and Force seem to merge into one another." What he did not realize was that through this means X-rays were also being generated by his apparatus (Crookes 1879, 130-31).

Crookes's conviction that the cathode rays were streams of particles carrying an electric charge was strenuously opposed by Goldstein, who reasserted the objection that if these rays really consisted of electrified particles, they should be subject to the influence of an electric field. Goldstein was unable to demonstrate any such influence and so concluded that the emanations were disturbances in the ether. A number of eminent scientists took Goldstein's side against Crookes.

SHADOWGRAPHS

The controversy over the nature of cathode rays took a startling and unexpected turn in December 1895 with the discovery of new rays, which were called X-rays because of their mysterious nature, by Wilhelm Konrad Röntgen (1845–1923), professor of physics at Würtzburg in Bavaria:

> If the discharge of a large Ruhmkorff coil is passed through a Hittorf vacuum tube, or through a sufficiently evacuated Enard, Crookes or similar apparatus, and the tube is covered with a fairly closely fitting sheath of thin black cardboard, it is observed that in a completely darkened room a paper screen painted with barium platino-cyanide and brought into the neighborhood of the apparatus lights up brightly or fluoresces at each discharge, irrespective of whether the painted or the other side of the screen faces the discharge apparatus. (Röntgen 1896, 272)

Röntgen's first communication, dated December 1895 was titled "Über eine neue Art von Strahlen" (On a New Type of Rays). It appeared in the January issue of the *Physiological Society of Würzburg*. A second communication with the same title, dated March 1896, appeared in the March issue of the same journal. He was 50 years old when he made the discovery that gained for him in 1901 the first Nobel Prize for Physics. His lifelong interest was in the physics of solids, and not in gas discharge phenomena; indeed, a review of his 49 published papers reveals that none had addressed the subject of gas discharges. As one might expect, the discovery of X-rays was not the culmination of a long-standing research program but really a fortunate accident involving a Crookes tube. One of the electrodes of his apparatus, the cathode, gave off an electric discharge when heated by the current, and a stream of cathode rays passed to the anode. The rays themselves were invisible but the greenish glow of the glass tube indicated that they were being given off.

Crookes made an experiment, which Röntgen repeated, of placing a mica shield in the form of a maltese cross in the middle of the tube, between the two electrodes, to see if it would cast a shadow. It did indeed. The shadow of the cross appeared on the wall of the glass tube. But this did not explain the nature of the rays. Perhaps they were ultraviolet rays? A screen coated with a fluorescent material, potassium platinocyanide, happened to be lying nearly on a laboratory table. It lit up. Röntgen then put the tube in a box made of thin black cardboard. To make sure that no light came through the box, he switched on the current to his tube. No light came through the box but, to his surprise, he noticed a strange glow in the far corner of his laboratory bench. Thinking that the glow was a figment of his imagination, he turned the switch again. He saw the glow again. Drawing back the curtains of his laboratory window, he found that the glow had come from the small fluorescent screen that had been placed at the far end of the table. Röntgen knew that the cathode rays would make the screen glow, but he also knew that they could not penetrate cardboard. Even if there was a minute leak in the cardboard box, he knew that they could not penetrate more than an inch or two of air. The fluorescent screen indicated that the mysterious rays could travel through the glass of the tube, the cardboard box, and air. He reckoned that they must consist of some unknown kind of invisible light. If so, they must cast a shadow. Following a sudden impulse, he placed his hand in front of the screen. He received the shock of his life. What he saw was not the shadow of his hand but a skeleton of a hand. He could see his own bones, with the flesh and skin forming a faint, grayish fringe around them.

This fortunate chain of events had revealed to Röntgen an entirely new kind of radiation. It could penetrate human flesh, wood and cardboard, metal foil and fabric, but it was stopped by bones and stone, thick metal, and other material of high density. Röntgen also found that the rays affected photographic plates, so that what he had seen on the fluorescent screen could be photographed. It became clear to him that this was a new form of light, which was invisible to the naked eye and which never had been observed or recorded.

It is very likely that other researchers had accidentally stumbled across X-rays but had failed to follow up the clues. Crookes, for example, had noticed that photographic plates lying near his experimental tubes were frequently fogged. He even returned some of the plates to the manufacturer, complaining of their spoiled condition. Other scientists had observed a fluorescent glow in materials located near Crookes tubes, but failed to identify the nature of the fluorescence or realize its source, as did Röntgen.

Röntgen spent the next eight weeks carefully repeating his experiments, checking each item of equipment, each procedure, and every observation to be certain that nothing had been overlooked. It was only then that he prepared a short manuscript in the form of a preliminary report that he handed on December 28, 1895, to the president of the Physiological Society of Würtzburg, a comparatively obscure organization. In his report, Röntgen described his apparatus and the resulting X-rays, as he called them. He detailed a series of experiments that clarified the differences between these rays and the cathode rays already familiar to physicists. The former had far greater penetrating power and, unlike the latter, were unaffected by a very intense magnetic deflection. (It would be shown soon that electric fields did not bend them either.) If the source of X-ray radiation were to be considered as that spot on the discharge tube where the cathode rays struck the glass walls of the tube, then the shifting of the cathode ray by a magnet would also shift the X-ray source to the new terminus, or anticathode, of the cathode rays.

Röntgen then reviewed the various photographs taken with his apparatus to demonstrate the true "ray" character of the emanations. The most dramatic were photographs of his hand, showing the bony structure, a photograph made through his laboratory door that registered the varying thickness of stiles and panels and especially the streaked areas made by brushing on lead-based paint. One photograph showed the shadow of a wire wrapped around a wooden spool, another a shadow of weights set in a covered wooden box, another the needle and degree markings of a compass in an enclosed metal case. Finally, he insisted that X-rays were not ultraviolet rays because they were not refracted in passing from air into various substances, such as water, carbon disulfide, aluminum, rock salt, glass, zinc, and so on. Nor were the rays reflected by them. Furthermore, X-rays were not polarized, as were ultraviolet rays. He speculated that they might represent longitudinal rather than transverse vibrations in the ether and that this opened up the need for a new chain of experiments.

His first communication was translated and published in English, French, Italian, and Russian. Röntgen also dispatched copies of his early photographs to scientists. A sensation was caused by the announcement that an obscure German professor had discovered rays that could make invisible things visible, that could pass through clothes, skin, and flesh to cast the shadows of the bones themselves on a photographic plate. To put it in context, though Ford had built his first car in 1893, transportation was still in the horse-and-buggy era. Flying was still a pipe dream of mad inventors. The transmission of sound by wireless was only a year in the future, but it would be 20 years before the first broadcast of a wireless program for the public would go out over the airwaves. Cinema was a year old, thanks to the opening of the first "kinetoscope parlour" on Broadway by Thomas Alva Edison, which showed moving pictures of the peep-show variety.

Enthusiasm outweighed disapproval. The frenzy of excitement had no earlier parallel. Despite the fact that the experiment was difficult to repeat (English lead glass was much less suitable than soft German glass to excite and transmit the rays), more than 100 papers appeared about this discovery in the first year after its announcement in 1895. A thousand books, pamphlets, and articles appeared in the public press on the subject of X-rays within the first year of their existence. Anyone and everyone who could lay their hands on an induction coil and a gas discharge tube got involved, from professional physicists to physicians and a host of amateur scientists.

When the right technique had been elaborated, the "hardness" of the X-rays proved invaluable to physics, medicine, crystallography, metallurgy; their exceedingly short length carried them through many opaque bodies. Physicists were besieged by physicians who recognized the great utility of the new rays. Before the end of the nineteenth century, "radiography" was applied in diagnosing lesions of the skull, the heart, and the lungs, and in renal pathology. In 1897, a course in radiology was inaugurated in Paris. That year, opaque substances were first used to radiograph the digestive system. Soon photographs of the human fetus, a tubercular patient's lungs, the heart and other organs were published. Soon, too, the power of X-rays to destroy organic cells was noticed. Röntgen worked in a zinc cabinet with a lead plate on the side. This afforded him considerable protection, but this precaution was taken to filter out stray emanations, and not to prevent exposure to such powerful penetrating rays. Others were not so lucky. Many cases of severe skin burns and loss of hair were reported, but no one appreciated the real danger. Noting their depilatory effect, one enterprising Frenchman, M. Gaudoin of Dijon, offered to use X-rays to remove unwanted hair from women's faces. He had numerous clients.

The first major improvement on the original Röntgen apparatus was a device to permit direct observation of an object, such as a hand, rather than an image on a photographic plate. This device was first described by the Italian physicist E. Salvioni (1858–1920) in February 1896. By placing a hand before a screen treated with platinocyanide and shielding out extraneous light from the eyes, the bone structure of the hand could be examined even as it moved. A further improvement involved treating the screen with salt platinocyanide of potassium. This device with its many subsequent improvements was known as the "fluoroscope." Edison was one of the first to work at improving the device. Putting his research team to work, more than 8,000 compounds were examined to see which gave the clearest pictures. The most effective was calcium tungstate. Edison arranged for this substance to be shaped into screens. He then turned to the design of the fluoroscope and selected one of 180 different variations of tubes.

Already in May 1896, Edison had arranged a public demonstration of X-rays to be held in connection with his National Electrical Exhibition in New York. The size of the crowd took the organizers by surprise. Among the assistants at the apparatus, who was in charge of the induction coil, was Clarence M. Dally. The lengthy exposure to the virulent rays resulted in severe burns and finally caused his death in 1904. As a result of injuries to Dally, Edison discontinued his work on X-rays. Many of the early users of the fluoroscope used one hand to hold the screen and positioned the other hand between the screen and the tube to gauge the strength of the X-ray emanation. It took the eyes a long time to adjust to the subdued light, which created a hazard to the hand. All too often the results were burns, lesions, amputations, and death.

MATERIAL CARRIERS OF ELECTRICITY

Röntgen rejected the idea that the radiation was, like light, due to transverse vibrations of the ether. Influenced by the fashionable hypothesis that the cathode ray was occasioned by a wave motion, he tentatively suggested that it might be longitudinal vibrations.

Joseph John Thomson was chosen at the early age of 28 to be the third Cavendish professor in 1884 (following Maxwell and Lord Raleigh), and he brought the well-equipped Cavendish Laboratory to bear on these new discoveries of radiation.

His familiarity with experimental work was limited, but Thomson learned quickly and presided over a flourishing of experimental culture at the Cavendish. Supported by his administration and teaching, many important experiments on electromagnetism and atomic particles were performed and many outstanding physicists received their early training, including seven Nobel Prize winners and 27 Fellows of the Royal Society. Thomson took an active interest in the work of all the young researchers at the Cavendish, daily checking on their progress and often making suggestions for improvements.

He had been thinking and experimenting in the general field of electric conduction in gases for many years. In 1882, he had published *Recent Researches in Electricity and Magnetism,* which contained a long chapter on the discharge of electricity through gases, the first detailed account of the subject in English, and the standard reference until it was followed by his *Conduction of Electricity Through Gases.* These two books were separated by more than a decade; Röntgen's discovery transformed the entire subject from one that was punctuated by uncoordinated empirical work to one that was coordinated by a convincing theoretical scaffolding.

With respect to the rays, scientists were agreed on a number of points: They were deflected by a magnetic field, they produced thermal and mechanical effects on matter placed in their path, and they produced phosphorescence when they fell on certain substances. In 1895, Jean-Baptiste Perrin (1870-1942) advanced a striking argument in favor of Varley's contention that cathode rays were streams of charged particles. He focused the cathode rays on a small, insulated cylinder originally designed by Faraday, which collected charges and registered them on a dial. He showed that when the rays fell on the Faraday cylinder, they deposited a negative electric charge at the point of impact. This result seemed to indicate that negative electrified particles were coming from the cathode, but it failed to demonstrate that the electrified particles were identical to the cathode rays. It might be that these particles emitted from the cathode coincidentally with the cathode rays.

With the collaboration of Ernest Rutherford (1871–1937), who at the time was engaged in postgraduate work at the laboratory, Thomson analyzed the ionizing power of the new rays discovered by Röntgen. When the rays passed through a gas, they made it a temporary conductor of electricity. When they were passed through a gas and then cut off, the conductivity of the gas was found to persist for a time and then gradually to die away. Thomson and Rutherford also discovered that when a gas, made conducting by X-rays, was passed through glass wool or between two oppositely electrified plates, the conductivity disappeared, indicating that it was due to charged particles that were discharged by contact with the glass wool or one of the electrified plates.

Rutherford found that, in a conducting gas, the current was first proportional to the applied electromotive force, but as that force was raised, the current increased more slowly and finally reached a maximum or saturation value. Such experiments demonstrated that while ions were part of the ordinary and permanent constitution of a liquid electrolyte, they existed in gases only when X-rays or other ionizing agencies are acting. Left to themselves, the ions gradually recombined and disappeared.

In 1897, Thomson employed superior vacuum pumps to remove more of the gas from the tube than Goldstein had been able to do. This made it possible for him to shift the cathode rays with an electrostatic field (fig. 11.2). The rays, he showed, veered toward the positive pole. Hertz had attempted this experiment 14 years previously, without success. The pressure of his gas was too high, thereby masking the effect of the electric field.

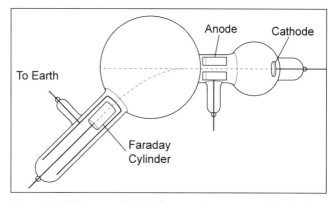

Figure 11.2: Diagram of J. J. Thompson's apparatus for showing that cathode rays are negatively charged. Artist: Jeff Dixon.

Thomson's method was afterward used by Rutherford to establish the nature of alpha rays; thus it will be useful to explain briefly how it works: An electric field exerts a force on a charged particle that depends upon the size of the charge only and is the same whether it is moving or at rest. The acceleration caused by the electric force, however, depends on the mass, so that the deflection, the moving aside, of a flying charged particle, will depend on the ratio of the charge, e, to the mass, m, usually written as e/m, and, of course, on the time for which the charge acts.

A magnetic field has no action of a charged particle at rest. A moving charged particle, however, is equivalent to an electric current: the faster the particle, the larger the corresponding current. A wire conveying an electric current in a magnetic field tends to move sideways, at right angles to itself and to the field. The movement of a beam of charged particles, then, in a magnetic field will depend upon the velocity as well as the ratio of charge to mass. Of course, in both an electric and magnetic field, the sideways movement of the beam in a given length of path will depend upon the velocity because that governs the time for which the particles are exposed to the field. However, this consideration is the same for both cases, though the velocity has an independent effect in the case of the magnetic field.

By measuring the deflection of the beam in an electric and in a magnetic field, both the ratio of the charge to the mass of the particles that make up the beam, and the velocity of the particles, can be found. Thomson found that the ratio of charge to mass was the same whatever the nature of the gas in the tube or of the metal of the electrode and that it was 770 times greater than it is for a charged hydrogen ion in the electrolysis of a liquid.

On April 30, 1897, Thomson advanced three startling new hypotheses about cathode rays based on his experiments:

1. Cathode rays are formed by negative charges of electricity transported by particles of matter that he called "corpuscles", having a mass about 1,000 times smaller than a hydrogen atom.[1] These material carriers of charge did not depend on the gas that was present in the cathode tube, and their mean free path did not depend on the density of the medium through which they traveled.
2. These corpuscles are constituents of the atom—they are a new state of matter in which the subdivision of matter is carried much further than what occurs in the common gaseous state, a state in which all matter was of a single type and was the substance of which all the chemical elements are made. Other people had measured the e/m ratio or suggested that the cathode rays were composed of particles, but Thomson was the first to say that the cathode ray was a building block of the atom. Consequently, the possibility arose of constructing models of the inner structure of atoms, starting from electrons and developing new and revolutionary explanation for the regularities that appeared in the table of elements.

The term *electron* had been coined in 1891 by G. Johnstone Stoney (1826–1911) to denote the unit of charge found in experiments that passed electric current through chemicals. It was this sense of the term that figured in the electron theory of Larmor and Lorentz. Unlike Thomson, however, Larmor and Lorentz did not describe the electron as a part of the atom, but rather as a structure in the ether. Indeed, when FitzGerald suggested in 1897 that Thomson's corpuscles were really "free electrons," he was actually *disagreeing* with Thomson's hypotheses. FitzGerald had in mind the kind of "electron" described by Larmor's theory. Thomson's proposal was risky, but he was proved right, and for his courage he is remembered as the discoverer of the electron.

3. These corpuscles are the only constituents of the atom. Thomson had leaped to the conclusion that the particles in the cathode ray (which we now call electrons) were a fundamental part of all matter.

Thomson's speculations met with some skepticism. The second and third hypotheses were especially controversial (the third hypothesis indeed turned out to be false). Years later he recalled, "At first there were very few who believed in the existence of these bodies smaller than atoms. I was even told long afterwards by a distinguished physicist who had been present at my lecture at the Royal Institution that he thought I had been 'pulling their legs'" (Thomson 1936, 34).

Gradually scientists accepted Thomson's first and second hypotheses, although with some subtle changes in their meaning. Experiments by Thomson, Lenard, and others through the crucial year of 1897 were not enough to settle the uncertainties. Real understanding required many more experiments over later years.

In the wake of Thomson's 1897 work, theories about the atom proliferated dedicated to answering one pressing question: If Thomson had found the single building block of all atoms, how could atoms be built up out of these corpuscles? Thomson proposed a model, sometimes called the "plum pudding" or "raisin cake" model, in which thousands of tiny negatively charged corpuscles swarm inside a sort of cloud of massless positive charge. As we shall see in the next and final chapter, this theory was struck down by Thomson's own former student, Ernest Rutherford. Using a different kind of particle beam, Rutherford found evidence that the atom has a small core, a nucleus. Rutherford suggested that the atom might resemble a tiny solar system, with a massive, positively charged center circled by only a few electrons. Later this nucleus was found to be built of new kinds of particles (protons and neutrons), much heavier than electrons.

NOTE

1. Instead of diminishing it, subsequent experiments performed by Thomson increased this ratio. In 1907, he found that the *e/m* for a cathode-ray corpuscle was 1,700 times as great as the *e/m* for a hydrogen ion.

12

THE ATOM AND THE NEW PHYSICS OF THE TWENTIETH CENTURY

RADIOACTIVITY

As important as Röntgen's substantive discovery of X-rays was to physics, his real legacy was that his work led directly to another field of research—radioactivity. It was known that X-rays produce marked effects on phosphorescent substances. Naturally, scientists wondered whether these or any other natural bodies produce anything like X-rays in turn. In 1896, Henri Becquerel (1852–1908), professor of physics at the Ecole Polytechnique in Paris, looked for a similar photographic effect of phosphorescence (i.e., the emission of light by a substance, which was not due to its being hot). Wanting to ascertain whether certain substances can retain light, as a hot poker retains heat, he exposed some salts to the rays of the sun and laid them afterwards on a photographic plate wrapped up in black paper to protect it from ordinary light. Among these were salts of the metal uranium. When the plate was developed, he found a beautiful silhouette of the crystals in black on the negative.

Becquerel reckoned that the sunlight caused the crystals to emit radiation, and so he erroneously believed that he had discovered a way of obtaining X-rays without going to the trouble of setting up a Crookes tube. However, this conjecture was put to the dustbin when, purely by chance, the Sun did not shine in Paris for several days. He had already set up his apparatus to perform a series of experiments and decided to develop the plates anyway, expecting the images to be very feeble. He found to his surprise that the same encrusted crystals placed with respect to the photographic plates in the same conditions and acting through the same screens, but kept in the dark, produced the same photographic effects. The silhouettes appeared with great intensity. What he found was that the uranium crystals were emitting rays independently of luminescence.

Becquerel tested his conjecture by repeating the entire experiment in a completely darkened room, which made no difference to the behavior of the crystal that still imprinted its image on the photographic plate. The mysterious rays, it appeared, emanated spontaneously from matter. Becquerel spent a great deal of time in an attempt to control the production of his new rays. He found that he was unable to affect the emission of rays from his uranium salts. He could heat them, cook them, compress them, in fact do whatever he liked to them, but whatever he did had no effect whatsoever on the rate at which the rays were produced or the amount that came out of his crystals. Nor could he in any way alter the ray-emitting properties of uranium by attacking it chemically.

In 1903, Marie Sklodowska Curie (1867–1934), inspired by Becquerel (her teacher and friend) came to the conclusion that the only explanation for the phenomenon must lie in the uranium atom because the other elements in the uranium salts had never by themselves shown any inclination to give off rays. Curie also studied compounds of thorium, next to uranium the element of highest atomic weight, and found that it sent out strong radiations similar to those of uranium. In concert with her husband, Pierre Curie (1859–1906), she elected to track down the source of these rays. Her apparatus consisted of two horizontal metal plates with a difference of potential of 100 volts between them. A layer of powdered substance was sprinkled on the lower one, and the passage of electric charge was taken with an electroscope. They set out producing compounds of uranium in their laboratory.

One source for uranium is the mineral pitchblende, a variety of uranium compound with the black, pitchlike luster of a freshly broken surface. The Curies worked with it and succeeded in preparing fairly pure uranium products. These, they believed, would give off a barrage of rays. The Curies were startled to find that the rays coming from their pure uranium compounds were not nearly as pronounced as those that came from the original pitchblende. Marie Curie rightly interpreted this negative result to indicate the presence of another element in the pitchblende that was a much more powerful source of rays than the uranium itself.

Whatever was producing the rays was present in the pitchblende in minute amounts. Undaunted, in a ramshackle shed that served as a laboratory, The Curies shifted through a ton of raw ore in order to pry out a fraction of a gram of the material she sought. After months of work, they had isolated a decigram of a strange new substance, one that was a million times more powerful than uranium as an emitter of rays. They had isolated the first sample of radium. It took several years for Marie Curie to separate out enough of the element to determine its atomic weight, which was determined to be 225.93. They also isolated polonium, named by Marie Curie after Poland.

The nature of the radiation remained to be determined. Marie Curie thought that at least two kinds of rays were being emitted. One kind could be detected by a magnetic field, whereas the other kind was unaffected and would only travel a few inches before disappearing. A further question was the nature of the source of the energy. Finally, it was not clear why a metal plate that had been close to, but not in contact with, samples of radium became radioactive themselves.

ALPHA AND BETA RAYS

Theorizing was not Marie Curie's strong point, and the mysteries of radioactivity were unraveled almost single-handedly by Ernest Rutherford. When Rutherford first began working on radioactivity in 1899, little was known about it apart from the result of the Curies that it was not limited to uranium alone but was also a property of thorium, radium, and polonium. It was also known that X-rays were complex, that is, that they comprised radiations with different powers of penetrating solid bodies, and Rutherford set about measuring the power of penetration of the uranium radiations, using ionization produced as a measure of the intensity of the rays. In his experiment, he used two large horizontal parallel plates, with a difference of potential between them. On the lower plate, he placed a layer of uranium compound that he covered with very thin metal foils, measuring with an electrometer the ionization produced with various numbers of foils.

Rutherford discovered that of the radiation from uranium one part was unable to pass through more than about the 50th of a millimeter of aluminum foil, while the other part would pass through about half a millimeter before its intensity was reduced by one-half. The first named, which Rutherford called alpha rays, are the barrage of alpha particles continually issuing from some radioactive substances. (The alpha particles are positively charged particles of matter made up of two protons and two neutrons, and so are identical to the nucleus of the helium atom, which is emitted from some radioactive substances.) They produce the most marked effects. The beta rays are the barrage of beta particles continually issuing from some radioactive substance. The beta rays are electrons that are emitted with great force from the nucleus of some radioactive substances. They are the more penetrating and can affect a photographic plate through opaque screens. In 1899, a third type of still-more-penetrating radiation, called gamma rays, was detected by the French chemist Paul-Ulrich Villard (1860–1934). These last rays can traverse plates of lead a centimeter thick. They are ultrashort X-rays of wave radiation that are emitted from the nucleus of some radioactive atoms. Rutherford reported his results in his "Uranium Radiation and the Electrical Conduction Produced by It" (1899).

Before this paper was published, he left the Cavendish for McGill University in December 1898. In his nine years at McGill, Rutherford's researches and those of his collaborators were entirely devoted to radioactivity. Rutherford's first major discovery was of a new class of radioactive substances, with peculiar properties. He found that thorium, besides emitting alpha and beta rays, gave out an active substance that could be carried from it by a current of air and seemed to be of the nature of a gas because it passed through cotton wool with the air, which particles of dust would not have done. Further proof was given by the fact that it could be condensed from the air by extreme cold, but was not affected by a high temperature. Both properties pointed to its being a gas. The radioactivity of the thorium gas or emanation was demonstrated by blowing a slow stream of air over it into a vessel with an insulated electrode at 100 volts. An electrometer indicated that it caused the leakage of electricity typical of ionized air.

As long as the flow of air was steady, the leak was steady. However, if air containing the thorium emanation was shut up in the vessel, a new phenomenon made itself evident. The leak due to radioactivity became slower and slower as time passed, as if the radioactivity was decreasing. Rutherford showed that this decrease was not due to ions being removed because if the emanation was merely allowed to stand, without the application of any potential difference, and so without any displacement of charged particles, it lost activity in exactly the same manner. The loss of activity was rapid. What's more, Rutherford found that, whatever its observed value at any moment, it lost half of its activity in 54 seconds (the half-life). The activity of thorium and uranium, as far as Rutherford knew, did not change as time went on. This activity decreasing with time was something entirely new. We now know that the activity of all radioactive substances decays, but in the case of thorium and uranium the half-life ran into hundreds of millions of years.

In 1900, after failing to secure a position at the University of Toronto, Frederick Soddy (1877-1956) accepted a junior post in chemistry at McGill. Collaboration with Soddy was critical to Rutherford's work on radioactivity. Rutherford was an expert in the measurement of radioactivity by ionization methods, and in the design and manipulation of apparatus for this purpose, but he had no serious training in chemistry, a detailed knowledge of which was essential for the separation of the radioactive elements. Soddy was an expert chemist with no experience of radioactive measurements. Both were enthusiastic about experiment and the atomistic hypothesis.

Their first collaboration showed that a very active substance could be separated from thorium by a simple chemical manipulation. Crookes had already separated from uranium an active substance that he called uranium X, so they called the new active substance thorium X. They found that the major part of the activity of thorium could be removed by precipitation with ammonia, and Soddy found that some elements—as it turns out, most elements—really exist in two or more varieties, identical in all their chemical and most of their physical properties and with the same atomic number (i.e., the same number of protons and electrons), but different in atomic weight. He called the variations isotopes (from the Greek *isos*, the same, and *topos*, place). Investigating the cause of uranium's radioactive rays, Soddy found that "ordinary" uranium, with an atomic weight of 238, was always mixed with small quantities of a lighter variety, with an atomic weight of 235, which was in a very unstable state. Like radium, it was breaking up, giving off the radiation that Becquerel had first observed on his photographic plates.

In the fall of 1903, Rutherford presented a new hypothesis. He noted that whenever there is radioactivity, there is (a) a chemical change, with new bodies appearing; (b) a disassociation of single particles and not a combination; (c) the activity is proportional to the mass of the radioactive element whether free or combined, so that the disassociating particles are atoms and not molecules; and (d) the amount of energy liberated is many thousand times more than that associated with the most violent chemical reaction. Rutherford explained these facts by the disintegration hypothesis—namely, radioactivity is due to an explosive disintegration of the elementary atom. Here and there one atom out of millions suddenly explodes; an alpha particle, or a beta particle and a gamma ray, are ejected, and a different kind of atom is left behind. What he proposed in effect was that the atom is not indestructible; large radioactive atoms could break themselves down to form smaller ones. Elaborating on this theory of atomic disintegration, Rutherford claimed that radioactivity was a phenomenon in which new types of matter are produced; that is, it is a change occurring within the atom. In short, the atom splits apart. These views were summarized in his *Radio-Activity*, which appeared in 1904.

Rutherford's view of the transmutation of the elements was endorsed by Oliver Lodge, then a leading scientist, but it was strongly opposed by William Thomson, who had become the senior figure in British physics and was now known as Lord Kelvin. He argued that the production of helium from radium no more proves transmutation than does the discovery of helium in clevite prove transmutation. He postulated instead that waves in the ether supplied radium with energy. (Kelvin also regarded the gamma rays as vapor from radium.) It was difficult for scientists from the old school to give up the belief in the unsplittable atom.

The portrait of the atom as drawn by Rutherford was the culmination of all these innovations. In Rutherford's model, the atom consisted of a central core, or nucleus, carrying a positive charge, and revolving around it in various orbits, a number of negatively charged electrons. As the electrons have very little mass, the whole mass of the atom must be concentrated in the nucleus. The nucleus is not a single, solid piece of matter but consists of protons, each with a positive electric charge equal to the negative one of an electron. Therefore, there must be the same number of protons in the nucleus as there are electrons revolving around it like tiny planets around a miniature sun, so that in its normal state the whole atom is electrically neutral. These electrical forces hold the atom together, but if an atom loses one or more electrons, or acquires additional ones, its electrical balance is disturbed—it becomes an ion. Thus X-rays and radium rays ionize their air by knocking electrons off its atoms; the alpha particles turned out to be ions of the gas helium, nuclei without their electrons.

Rutherford did not assume that the electron and the proton were the only kinds of particles to be found in the atom. In 1920, he demurred that an electrically neutral particle may yet be uncovered, which he visualized as a close combination of a positively charged proton and a negatively charged electron so that the whole particle would have no electrical charge. He said that if such a particle existed, it would move freely through matter, being neither repelled nor attracted by other particles.

THE CHARGE OF AN ELECTRON

Between 1908 and 1917, a series of remarkable experiments was carried out by the American physicist Robert Andrews Millikan (1868–1953), an assiduous experimenter with a penchant for contributing to the hot topics of the day. These experiments, which were designed to obtain an accurate value of the charge carried by an electron, played a crucial role in the acceptance of the existence of the electron.

The apparatus associated with Millikan's oil-drop experiment is depicted in figure 12.1. A sealed chamber with transparent sides was fitted with two parallel metal plates, which acquired a positive or negative charge when an electric current was applied by a bank of batteries. A perfume atomizer sprayed a fine mist of oil droplets into the upper portion of the chamber. Under the influence of gravity and air resistance, some of the oil droplets fell through a small hole cut in the top metal plate, at which point they were illuminated by an arc lamp. When the space between the metal plates was ionized by X-rays, electrons from the air attached themselves to the falling oil droplets, causing them to acquire a negative charge. A light source, set at right angles to a viewing microscope, illuminated the oil droplets and made them appear as bright stars while they fell.

With this equipment, the mass of a single charged droplet could be calculated by observing how fast it fell. By adjusting

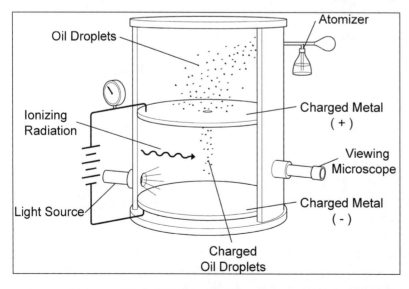

Figure 12.1: Diagram of R. A. Millikan's oil-drop experiment. Artist: Jeff Dixon.

the voltage between the metal plates, the speed of the droplet's motion could be increased or decreased; when the amount of upward electric force equaled the known downward gravitational force, the charged droplet remained stationary. The amount of voltage needed to suspend a droplet was used along with its mass to determine the overall electric charge on the droplet.

From the time of Millikan's original experiment, this method offered convincing proof that electric charge exists in basic natural units. All subsequent distinct methods of measuring the basic unit of electric charge point to its having the same fundamental value.

FROM BOHR TO QUANTUM ELECTRODYNAMICS

In 1912, the Danish physicist Niels Bohr (1895–1962) joined Rutherford at Manchester in 1912. Bohr realized at once that, while radioactivity was bound up with the nucleus, the ordinary chemical and physical properties of matter must be an affair of the system of electrons surrounding the nucleus. He proceeded to show how these properties could be explained.

In so doing, he disarmed the one pressing objection to Rutherford planetary model of the atom. Rutherford's model, with electrons orbiting a central nucleus, was theoretically unstable. Unlike planets orbiting the Sun, electrons are charged particles. They should radiate energy and consequently spiral in toward the nucleus. In three classic papers published in 1913, Bohr elaborated a new quantum theory of the atom.

Bohr's scheme was based on the discovery of the German physicist Max Planck (1858–1947) that radiation, such as light, was emitted not continuously, as seemed natural, but in little packets of radiant energy, which he called *quanta*. The size of the quantum of radiant energy depends upon the frequency of the radiation in question. In fact, the size of the quantum energy of a particular radiation is the frequency of the radiation multiplied by a small number, always denoted by h, called Planck's constant.

When this conception of radiant energy in packets, the size of which is proportional to the frequency, was first announced by Planck in 1901, it attracted little attention. An exception was a remarkable paper (Einstein 1905) published in 1905 by Albert Einstein (1879–1955), which extended Planck's insight that hot objects emit electromagnetic radiation in discrete packets, in direct contradiction with the classical view of light as a continuous wave. Although Einstein's theory of the photoelectric effect had shown that the quantum theory could explain fundamental observations in the fields of photoelectricity (the release of electrons by light) and specific heat, the notion of a quantum was not widely accepted until Bohr's work. The new guiding assumption of Bohr's work, namely, the laws of electromagnetism that explained large-scale events, such as the behavior of electric waves used in wireless telegraphy, did not apply to atoms, which had rules of their own.

Bohr adopted Rutherford's nuclear atom, considering the nucleus to be surrounded by electrons circulating in orbits, like planets round the Sun, and governed by the ordinary laws of electrical attraction, just as planetary orbits are governed by gravitational attraction. However, in opposition to the classical laws of electromagnetism, he asserted that (a) electrons that circulate in these orbits do not radiate; (b) of all the indefinite variety of orbits that were permitted by the classical laws, only certain widely separated ones were actually possible, these possible ones being determined by a special quantum condition; and (c) the electron absorbs or emits radiation only when it passed from one stationary state (or orbit) to another—absorbing energy as it moves away from the nucleus and emitting energy as it jumps back to the nucleus.

By the time of the appearance of Bohr's new model of the atom, it was clear that electrical and magnetic phenomena were bound up with the atom. Although the model furnished a powerful new model of the atom, it was not robust enough to accommodate heavier atoms with many electrons. Indeed, the German physicist Werner Heisenberg submitted in 1925 that the behavior of the parts of the atom are so complex that their positions in space and time cannot be determined with precision; in short, any attempt to determine the position of an electron invariably disturbs it, so that the best scientists can hope for is a probabilistic account. A year after the statement of this uncertainty principle, physicists were led to embrace the strange duality proposed by the French physicist Louis de Broglie (1887–1961) and given mathematical form by the Austrian physicist Erwin Schrödinger (1887–1961) that the energy of an electron behaves simultaneously like particles and waves. This model, in turn, was followed by the startling claim of the English physicist Paul Dirac (1902–1984) that there are subatomic particles that possess charges opposite to the electron and proton. The discovery of the positron in 1932 was a triumph for Dirac's extension of quantum mechanics to the study of electrons moving at speeds approaching that of light.

Quantum mechanics also led to the creation of a quantum theory of electromagnetism called quantum electrodynamics, or QED. Completed in the 1940s, when experiments conducted during the Second World War that used microwave technology stimulated new avenues of investigation, QED addresses the processes involved in the creation of elementary particles from electromagnetic energy, and the reverse processes in which a particle and its antiparticle destroy each other and produce energy. The fundamental equations of QED apply to the emission and absorption of light by atoms and the basic interactions of light with electrons and other elementary particles. Charged particles interact by emitting and absorbing photons, the particles of light that transmit electromagnetic forces. For this reason, QED is also known as the quantum theory of light.

TIMELINE

c. 580 B.C.	Thales studies amber and the lodestone.
c. 450 B.C.	Empedocles develops the effluvial theory of electricity.
c. 400 B.C.	Democritus expounds theory that all things are made up of tiny, indivisible particles or atoms.
56 B.C.	Lucretius composes *De rerum natura*.
A.D. 413–436	Augustine reports on the phenomenon of magnetic induction.
1269	Peregrinus composes "Epistle on the Magnet."
1287	Roger Bacon pioneers the experimental method.
1455	Johann Gutenberg publishes the Bible, the first European printed book set in movable types.
1492	Columbus sails west from Spain, notes declination changes in mid-ocean from easterly to westerly.
1514	Nicolas Copernicus presents the central ideas of his heliocentric theory in *Commentariolus*.
1543	Copernicus's heliocentric system is presented in *On the Revolutions of the Heavenly Spheres*.
1558	Peregrinus's letter appears in book form.
1576	Thomas Digges modifies the Copernican system by removing its outer edge and replacing the edge with a star-filled, unbounded space.
1581	Robert Norman publishes "The Newe Attractive," with its announcement of magnetic dip (inclination).
1600	William Gilbert publishes *On the Magnet*.
1604	Galileo discovers the laws of pendulum and free fall by experiment.
	Kepler suggests that light obeys an inverse square law.

Timeline

1605	Francis Bacon publishes *Advancement of Learning*.
1609	Kepler publishes *Astronomia nova*, containing a physical explanation for planetary motion and the first two laws that bear his name.
	The first Accademia dei Lincei (listing Galileo as a member) is created in Rome.
1638	Descartes introduces the concept of an ether filling all of space to explain the tides.
1644	Descartes publishes *Principles of Philosophy*, containing the vortex theory of planetary motion and arguments concerning the impossibility of vacua.
1650	Otto von Guericke constructs the first vacuum pump (published in 1672).
	Harvard University founded.
	First chemistry laboratory established at Leyden University.
1651	Thomas Hobbes publishes *Leviathan*, elaborating a materialistic doctrine of nature and human society.
	Guericke demonstrates pressure of atmosphere using "Magdeburg hemispheres."
1660	Royal Society of London founded.
1665	Isaac Newton deduces the inverse-square gravitational force law from the "falling" of the moon.
	Newton invents his calculus.
	Royal Society secretary begins *Philosophical Transactions*, the first journal of a strictly scientific nature.
1666	Académie des Sciences founded.
1671	Guericke describes electric conduction and repulsion in his *Experimenta Nova*.
1675	Robert Boyle publishes first book on electricity: *Experiments and Notes About the Mechanical Origine or Production of Electricity*.
1687	Newton publishes *Principia Mathematica*.
1700	On the advise of Leibniz, Frederick I created the Société des Sciences in Berlin.
1699	Edmund Halley conducts the first magnetic survey.
1704	Newton publishes *Opticks*.
1727	American scientists meet, at Benjamin Franklin's initiative, and form the Leathernapron Club, the nucleus for the American Philosophical Society.
1729	Stephen Gray identifies differences between insulators and conductors.
1730	Approximate date for commencement of Enlightenment period.

1733	Charles François Dufay advances two-fluid theory of electricity.
1745	Pieter van Musschenbroek and Ewald G. von Kleist discover the principle of the Leyden jar, in which static electricity charges could be stored (announced by Musschenbroek in 1746).
1747–1754	Franklin elaborates single-fluid theory of electricity.
1750	John Michell publishes *A Treatise of Artificial Magnets*, the first attempt to establish the inverse square law of force between magnetic poles.
1751–1772	Denis Diderot and Jean le Rond d'Alembert coedit the *Grande Encyclopédie*, widely regarded as the most influential work of the eighteenth century.
1752	Franklin shows that lightning is electricity.
1753	Founding of British Museum, around the library and collection of Hans Sloane.
1759	Aepinus (Frans Ulrich Theodosius) publishes *An Attempt at a Theory of Electricity and Magnetism*.
1767	Joseph Priestly proposes an electrical inverse-square law.
1772	Antoine Lavoisier experiments on combustion.
1781	Lavoisier states law of conservation of mass.
	Immanuel Kant publishes *The Critique of Pure Reason*, which asserts the impossibility of knowledge regarding the existence of atoms.
1785	Charles Coulomb announces his inverse square law of electrical attraction and repulsion.
1786	Luigi Galvani discovers animal electricity and postulates that animal bodies are storehouses of electricity (published in 1789).
	Kant publishes *The Metaphysical Foundations of Natural Science*, which sets the stage for the emergence of *Naturphilosophie*.
1789	Declaration of Rights of Man and of the Citizen passed in France.
1791	Galvani publishes his findings in electrophysiology (experiments carried out in 1789).
1799	Alessandro Volta invents the electric battery (announced in 1800).
	The Royal Institution is founded in London.
1800	William Nicholson discovers electrolysis.
	Nicholson and Anthony Carlisle use electrolysis to separate water into hydrogen and oxygen.
1801	William Hyde Wollaston shows that frictional and galvanic electricity are the same.

	Founding of the Royal Institution (Great Britain).
1803	Dalton advances atomic theory, contending that matter is composed of discrete particles of different weights.
1806	Davy's first Bakerian lecture, "On Some Chemical Agencies of Electricity."
1807	Davy discovers potassium.
1808	Dalton's atomic theory of chemical combinations laid out in *A New System of Chemical Philosophy*.
	Davy develops the first electric powered lamp.
1812	Poisson puts the finishing touches on the mathematical apparatus of Coulomb's electrodynamcs.
1813	Hans Christian Oersted suggests that electricity ought to be convertible to magnetism.
	Publication of the French edition of Ptolemy's *Almagest*, prepared by Nicolas Halma, marks the beginning of the critical publication of classical works in the history of science.
1819	Oersted shows that an electric current is able to deflect a magnetic needle (published in 1820).
	Dominique François Arago shows that a current can magnetize iron (the electromagnet).
	André Marie Ampère commences work on electromagnetism.
1821	Michael Faraday shows that electrical forces can produce rotational motion of a needle.
	Thomas Johann Seebeck converts heat into electricity.
1824	Poisson publishes a mathematical treatment of magnetism to complement the one that he had published for electrostatics in 1812.
1825	Ampère deduces law for force between current-carrying conductors.
	William Sturgeon, a British electrician, constructs an electromagnet.
1826	Georg Simon Ohm proposes what is now called Ohm's law relating current and voltage.
	Faraday founds the popular lecture series for children called the Christmas Lecture Series.
1829	Joseph Henry invents the first practical electric motor.
1831	Faraday discovers electromagnetic induction and devises the first electrical generator. Begins work on electrolysis.
	Joseph Henry independently obtains similar results in the United States.
	Joseph Henry proposes an electromagnetic telegraph.

	The British Association for the Advancement of Science is formed.
1832	Samuel Morse invents the telegraph and a telegraphic alphabet.
	Faraday introduces the terms *electrode, anode, cathode,* and *ion.*
1833	Faraday announces the laws of electrolysis.
	At a meeting of the British Association for the Advancement of Science, William Whewell proposes the term *scientist.*
	Faraday appointed Fullerian professor of Chemistry at the Royal Institution.
1835	Morse produces the first working model of the electric telegraph.
1837	Charles Wheatstone and James Cooke produce the first practical telegraph system.
1843	Morse builds the first long-distance electric telegraph line.
1844	Morse patents his design for the telegraph.
1845	Faraday discovers effect of magnetic field on polarized light (Faraday effect).
1846	The Smithsonian Institution is founded in the United States.
	In the short paper "Thoughts on Ray Vibrations," Faraday discusses the possibility of doing away with the ether.
1848	Establishment of the American Association for the Advancement of Science.
1852	Faraday publishes "On the Physical Character of the Lines of Magnetic Force," charging that Newtonian science is an obstacle to the progress of science.
1863	The National Academy of Sciences is founded in the United States.
1866	Leclanché invents the electrical battery (dry cell).
1871	Eugen Goldstein submits that cathode radiation sustains a wave interpretation of matter.
	Cromwell Varley submits that cathode radiation sustains a particulate theory.
1872	Cavendish Laboratory at Cambridge completed after four years of construction.
1873	James Clerk Maxwell publishes his *Treatise on Electricity and Magnetism,* fully develops the view that light is an electromagnetic phenomenon (first published in two memoirs of 1865 and 1868).

1875	Crookes claims that cathode radiation points to a fourth state of matter that is particulate in nature.
1876	Alexander Graham Bell patents the telephone.
	Julius Plücker describes the phenomenon of cathode rays and is the first to use the term.
1877	Thomas Edison invents the phonograph.
1879	Albert Abraham Michelson determines the velocity of light.
	Edison patents the incandescent light bulb.
1882	The Pearl Street Power Station brings electricity to New York City.
1887	Michelson and Edward Morley attempted to measure the changes in the velocity of light produced by the motion of Earth through space.
	Heinrich Hertz observes the photoelectric effect.
1888	Hertz demonstrates propagation of electromagnetic waves (radio waves).
	Eastman markets his handheld box camera.
	Nikola Tesla patents the induction motor.
1891	George Johnstone Stoney suggests the term *electron* for unit of electricity.
1892–1904	Hendrik Antoon Lorentz develops his electron theory.
1894	Guglielmo Marconi begins experimenting on communicating with radio waves.
1895	Wilhelm Konrad Röntgen discovers X-rays, which were immediately used to visualize bodily structures.
1896	Henri Becquerel discovers natural radioactivity in uranium salts.
	Marconi patents radio telegraphy.
1897	Thomson shows that electrons are independent particles. Announces the existence of this fundamental particle during a lecture at the Royal Institution.
	Inauguration of first course in radiology.
1898	Marie and Pierre Curie isolate the radioactive elements of radium and polonium.
	John Sealy Townsend measures the charge of an electron.
1899	Alpha and beta rays shown by Ernest Rutherford to be distinct types of radiation.
	Rutherford notices thorium emanation.
1900	Max Planck states his radiation law and Planck's constant.
1902	Rutherford and Frederick Soddy discover the law of radioactive decay.

1903	Rutherford and Soddy publish work on radioactivity as the transformation of atoms.
1905	Albert Einstein completes his theory of special relativity.
	Einstein explains the photoelectric effect.
1908	Hans Geiger and Rutherford invent the Geiger counter.
1909	Rutherford and Thomas D. Royds demonstrate that alpha particles are doubly ionized helium atoms.
1911–1913	Soddy, Russell, and Kazimir Fajans recognize the existence of isotopes among radioactive elements.
	Rutherford proposes nuclear model of the atom.
1913	Niels Bohr presents his first quantum model of the atom (*On the Constitution of Atoms and Molecules*).
	Robert Millikan measures the fundamental unit of charge.
1926	Baird transmits the first television signal.
1927	Werner Heisenberg states the quantum uncertainty principle.
	Bohr formulates complementarity principle.
1932	Chadwick discovers the neutron.
1946	John Eckert and John Mauchley finish ENIAC, the first electronic computer.

GLOSSARY

action at a distance: The interaction of two objects that are separated in space with no known mediator of the interaction.

alpha particle: One of three types of radiation produced by naturally radioactivite substances identified and named by Rutherford in 1909. By measuring the charge and mass of alpha particles, he found that they are the nuclei of ordinary helium atoms (two protons and two neutrons). See beta particle.

ampere: The practical unit of electric current flow. If a one-ohm resistance is connected to a one-volt source, one ampere will flow.

Ampère's law: Named after its discoverer, this law relates the circulating magnetic field in a closed loop to the electric current passing through the loop. It is the magnetic equivalent of Faraday's law of induction.

anode: The positive electrode of a device in an electric circuit. See Cathode.

astatic needle: A magnetic needle, developed by Ampère, that is devoid of polarity, and so does not have a tendency to point in a given direction.

beta particle: One of three types of radiation produced by naturally radioactive substances identified and named by Rutherford, who found that it consists of high-speed electrons. See alpha particle.

capacitor: An electronic device that stores energy in an electric field by accumulating an internal imbalance of electric charge. The principle of the capacitor was discovered independently in 1745 by Pieter van Musschenbroek and Ewald G. von Kleist.

cathode: The negative electrode of a device in an electric circuit. In a vacuum tube, electrons flow from the anode toward the cathode. See anode.

cathode tube: A vacuum tube in which a hot cathode emits electrons that are accelerated as a beam through a relatively high-voltage anode.

charge: A basic property of matter that gives rise to all electric and magnetic forces and interactions. Matter can be "neutral" (having no electrical charge), or it can have one of two kinds of charges distinguished as "positive" or "negative." See coulomb.

chemical affinity: The tendency of certain atoms or molecules to aggregate or bond.

chemical reaction: A process by which atoms of the same or different substances rearrange themselves to form a new substance. During this process, they either absorb heat or give it off.

condenser: An antiquated name for a capacitor.

conductor: An electrical path that presents comparatively little resistance.

coulomb: A unit of electrical charge equal to the quantity of electricity passing in one second through a circuit in which the rate of flow is one ampere.

Coulomb's law: The fundamental law of electrostatics, which states that the force between two charged particles is directly proportional to the product of their charges and inversely proportional to the square of the distance between them.

diamagnetism: A weak form of magnetism that is occasioned by changes in the orbital motion of electrons due to the presence of an external magnetic field.

dielectric: An insulator (e.g., glass, rubber, plastic). Dielectric materials can be made to hold an electrostatic charge, but current cannot flow through them.

disintegration hypothesis: Rutherford's hypothesis that radioactivity is due to an explosive disintegration of the elementary atom.

displacement current: A hypothetical current proposed by Maxwell when formulating what are today known as Maxwell's equations.

dualistic system: A classification of chemical elements into one group that is electrically positive and another group that is negative, which combine to form neutral products that can be decomposed by an electric current.

effluvia: An invisible emanation or exhalation, as of vapor or gas, that became the basis for the theory of electricity associated with William Gilbert.

electric battery: A device invented by Volta that converts chemical energy into electrical energy, consisting of a group of electric cells that are connected to act as a source of continuous current.

electric circuit: The unbroken path along which an electric current exists or is able to flow.

electric current: The movement of electrons through a conductor, measured in amperes, milliamperes, and microamperes.

electric potential: The difference in electrical charge between two points in a circuit expressed in volts. See electrical tension.

electrical field: A region of space, associated with a distribution of electric charge, in which forces due to that charge act upon other electric charges.

electrical fluid(s): A theoretical entity(ies) invoked to explain electrical phenomena. One- and two-fluid theories of electricity reached their apex at the end of the eighteenth century.

electrical tension: An early term for electrical potential.

electrician: A person who is versed in the knowledge of electricity.

electricity: A cluster of phenomena arising from the behavior of electrons and protons that is caused by the attraction of particles with opposite charges and the repulsion of particles with the same charge.

electricity (animal): The electricity developed in some animals, such as the electric eel, torpedo, and others.

electricity (resinous): The electricity excited by rubbing amber with cat's fur; that is, negative electricity, as opposed to vitreous or positive electricity. Bodies electrified (charged) with vitreous electricity attract bodies electrified with resinous electricity; and repel other bodies electrified with vitreous electricity. See electricity (vitreous).

electricity (vitreous): The electricity excited by rubbing glass with such substances as silk; that is, positive electricity, as opposed to resinous or negative electricity. See electricity (resinous).

electricity (voltaic): Direct current electricity, notably as produced by chemical action.

electrode: The two electronically conducting parts of an electrochemical cell.

electrodynamics: Classical electrodynamics (or classical electromagnetism) is a theory of electromagnetism that was developed over the course of the nineteenth century, most prominently by Maxwell. It provides a robust description of electromagnetic phenomena whenever the relevant length scales and field strengths are large enough that quantum mechanical effects are inappreciable.

electrolysis: Chemical change, especially decomposition, produced in an electrolyte by an electric current.

electromagnet: A magnet produced by passing an electric current through an insulated wire conductor coiled around a core of soft iron, as in the fields of a dynamo or motor.

electromagnetic field: A field of force associated with a moving electric charge, consisting of electric and magnetic fields that are generated at right angles to each other.

electromagnetic induction: Production of an electric current by changing the magnetic field enclosed by an electrical circuit. This relation was discovered by Michael Faraday in 1831.

electromagnetic radiation: The manifestation of the electromagnetic interaction between charged particles.

electromagnetic waves: A changing electromagnetic field propagates away from its origin in the form of a wave. These waves travel in vacuum at the speed of light and exist in a wide spectrum of wavelengths.

electromagnetism: The physics of the electromagnetic field, encompassing all of space, which exerts a force on those particles that possess the property of electric charge, and is in turn affected by the presence and motion of such particles.

electrometer: A voltmeter with large input resistance

electromotive force (EMF): An energy-charge relation that results in electric pressure (voltage), which produces or tends to produce charge flow.

electron: A basic constituent of matter carrying a unit charge of negative electricity. Electric current is the flow of electrons through a wire conductor (*see* electricity). The name *electron* was first used for a unit of negative electricity by the English physicist G. J. Stoney, but the particle was identified as such in 1897 by J. J. Thomson, who showed that cathode rays are composed of electrons (which he called "corpuscles") and who measured the ratio of charge to mass for the electron.

electrophorus: A device invented by Volta that is used to produce a large static electric charge by induction.

electrotonic state: A hypothetical state of matter advanced by Faraday whereby the particles of matter are in a state of tension due to the action of an electric current.

ether: A substance that fills all of space and propagates electromagnetic radiation. The concept fell into disfavor after the publication of Einstein's theory of relativity.

Faraday rotation: The discovery made by Faraday that a wave of light polarized in a certain plane can be turned about by the influence of a magnet so that the vibrations occur in a different plane.

Faraday's laws: Two fundamental laws of electrochemistry discovered by Faraday: (1) in any electrolytic process, the amount of chemical change produced is proportional of the total amount of electrical charge passed through the cell; and (2) the mass of the chemicals changed is proportional to the chemicals' equivalent weight.

fluorescence: That property by virtue of which certain solids and fluids become luminous under the influence of radiant energy.

fluoroscope: An imaging device that enables physicians to obtain real-time images of organic structures.

force: Something that causes a change in the motion of an object. The modern definition of force (an object's mass multiplied by its acceleration) was given by Isaac Newton in 1687.

Fresnel drag coefficient (also called the Fresnel ether-drag hypothesis): In 1818 Fresnel proposed that substantial material bodies might carry some of the ether along with them. The Fresnel drag coefficient is

$$(1 - \frac{1}{n^2})$$

where n is the refractive index of the body.

galvanometer: A device that measures the strength of electric currents by means of the deflection of a magnetic needle, around which a current is caused to flow through a coil of wire.

gamma ray: Electromagnetic radiation emitted from the nucleus of an atom by radioactive decay and having energies in a range from ten thousand (10^4) to ten million (10^7) electron volts.

generator: A general name given to any device that transforms mechanical energy into electrical energy.

half-life: The time in which one-half of the atoms of a particular radioactive substance disintegrates into another nuclear form.

induction: The process by which an electrical conductor becomes electrified when near a charged body and becomes magnetized.

invisible college: First used by Boyle, the expression "the invisible college" now refers to an extensive network of scientists, their colleagues, associates, and research collaborators, which keeps scientists abreast of the latest developments and trends in their fields of research.

insulator: A device for fastening and supporting a conductor.

ion: An atom, molecule, or molecule fragment that has acquired a net electric charge by gaining or losing one or more electrons. *Anions* contain excess electrons and are negatively charged, and *cations* are deficient in electrons and are positively charged.

leakage: The escape of electric current through defects in insulation or other causes.

Leyden jar: An early form of capacitor consisting of a glass jar lined inside and out with tinfoil and having a conducting rod that connected to the inner foil lining and passed out of the jar through an insulated stopper.

magnet: An object that attracts iron and some other materials. Magnets are said to generate a magnetic field around themselves. Every magnet has two poles, called the north and south poles. Magnetic poles exert forces on each other in such a way that like poles repel and unlike poles attract each other.

magnet (permanent): A piece of magnetic material that retains its magnetism once it is removed from a magnetic field.

magnetic compass: A small magnet that is affected by the magnetic field of the Earth in such a way that, when mounted on a pivot, it points to a magnetic pole of the Earth.

magnetic field: A magnetic field is said to exist in a region if a force can be exerted on a magnet. If a compass needle is deflected when it is put at a particular location, we say a magnetic field exists at that point, and the strength of the field is measured by the strength of the force of the compass needle. All known magnetic fields are caused by the movement of electrical charges. Electrons in orbit in atoms give rise to magnetic fields, so that every atom is, like the Earth, surrounded by a magnetic field:

magnetism: Force of attraction or repulsion between various substances, especially those made of iron and certain other metals, which is due to the motion of electric charges.

mechanics: The branch of physics that addresses the action of forces on matter.

Maxwell's equations: The set of equations, attributed to Maxwell, that describes the behavior of the electric and magnetic fields, as well as the interactions of these fields with matter. Maxwell formulated these equations in 1865 in terms of 20 equations in 20 variables, including several equations now considered to be secondary to what are now called "Maxwell's equations." The modern mathematical formulation of Maxwell's equations is due to Heaviside, who reformulated Maxwell's original system of equations to a far simpler representation using vector calculus.

molecular vortex: Maxwell's model of electromagnetism, which assumes that every magnetic tube of force is a vortex with an axis of rotation coinciding with the direction of the force.

naturphilosophie: A philosophy of nature prevalent in German literature, philosophy, and science from 1790 to 1830 or thereabouts. One of the central pillars of this intellectual movement is the belief in the unity of nature and its forces.

nuclear physics: Branch of physics concerned with the components, structure, and behavior of the nucleus of the atom.

ohm: The unit of electrical resistance. Resistance is one ohm when a direct current voltage of one volt will send a current of one ampere through.

Ohm's law: Named after its discoverer, Georg Ohm, this law states that the direct current flowing in a conductor is directly proportional to the potential difference between its ends. It is usually formulated as $V = IR$, where V is the potential difference, or voltage, I is the current, and R is the resistance of the conductor.

optical interferometer: An optical instrument that transmits two radio signals or beams of light and uses the interference principle (how they reinforce or neutralize each other) to measure wavelengths and wave velocities, and to calculate indices of refraction.

paramagnetic: Material that are naturally magnetic or can be made magnetic with ease.

phosphorescence: The emission of light by a substance that is not due to its being hot.

photoelectric: Descriptive of the effect that light has on electric circuits, through a device controlled by light.

polarization of light: Polarization is a phenomenon proper to transverse waves, that is, to waves that vibrate in a direction perpendicular to their direction of propagation: Light is a transverse electromagnetic wave.

proton: An elementary particle having a single positive electrical charge and constituting the nucleus of the ordinary hydrogen atom.
quantum: An elemental unit of energy.
quantum electrodynamics (QED): Quantum field theory that describes the properties of electromagnetic radiation and its interaction with electrically charged matter in the framework of quantum theory.
radioactivity: The spontaneous disintegration of the atomic nuclei of a radioactive substance resulting in the emission of ionizing radiation.
ray: The line of propagation of waves from a source. The term is also applied to streams of particles, such as the electrons emitted from a cathode.
resistance: The opposition offered by a substance or body to the passage through it of an electric current, which converts electric energy into heat: *See* ohm
resistor: A resistor is a two-terminal electrical or electronic component that resists an electric current by producing a voltage drop between its terminals in accordance with Ohm's law.
solenoid: A spiral of conducting wire wound cylindrically so that when an electric current passes through it, its turns are nearly equivalent to a succession of parallel circuits, and it acquires magnetic properties similar to those of a bar magnet.
spark: A discharge of electricity—accompanied by heat and incandescence—across a gap between two electrodes. :
special theory of relativity: The physical theory published by Einstein in 1905, based on two assumptions: (1) the speed of light in a vacuum is a constant; and (2) the laws of physics are invariant in all inertial systems
static electricity: An electrical charge that accumulates on an object when it is rubbed against another object.
transformer: A device, which consists of primary winding, secondary winding, and an iron core, that is used for changing the voltage and current of an alternating circuit.
volt: The practical unit of electric pressure equal to the pressure that will produce a current of one ampere against a resistance of one ohm. This unit with its symbol "V" is named in honor of Alessandro Volta.
voltage: A term sometimes used interchangeably with electrical potential.
voltaic pile: The first (1800) laboratory source of electricity, essentially what we would call today a nonrechargeable battery. Volta assembled (piled up) different metal disks (e.g., silver and zinc) separated by solution-soaked pasteboards, repeating the pattern many tens of times.

BIBLIOGRAPHY

PRIMARY SOURCES

Aepinus, Franz. 1759. *Tentamen Theoriae Electricitatis et Magnetismi.* St. Petersburg: Typis Academiae Scientiarum.

Ampère, André-Marie. 1820. "Sur l'action mutuelle entre deux courans électrique, entre un courant électrique et un aiment ou le globe terrestre, et entre deux aimans." *Annals de Chimie et de Physique* 15: 59–76, 170-208.

Ampère, André-Marie. 1827. *Mémoire sur la théorie mathématique des phénomènes électrodynamiques uniquement déduite de l'expérience.* In Académie des sciences, Institut de France (Ed.), *Mémoires de l'Académie des sciences de l'Institut de France, tome. 6, année 1823.* pp. 175-388

Bacon, Francis. 1605. *The tvvoo bookes of Francis Bacon: Of the proficience and aduancement of learning, diuine and humane.* Printed by T. Purfoot for H. Tomes.

Barlowe, William. 1597. *The Navigator's Supply.* London: G. Bishop, R. Newbery, and R. Barker. Reprinted in Amsterdam by Theatrum Orbis Terrarum (1972).

Bohr, Niels. 1913a. "On the Constitution of Atoms and Molecules." (Part 1 of 3) *Philosophical Magazine* 26: 1–25.

Bohr, Niels. 1913b. "On the Constitution of Atoms and Molecules, Part II Systems Containing Only a Single Nucleus." (Part 2 of 3) *Philosophical Magazine* 26: 476-502.

Bohr, Niels. 1913c. "On the Constitution of Atoms and Molecules, Part III." (Part 3 of 3) *Philosophical Magazine* 26: 857-875

Boyle, Robert. 1675. *Experiments and Notes About the Mechanical Origine or Production of Electricity.* London: E. Flesher for R. Davis.

Conton, John. 1754. "A Letter ... concerning some new electrical experiments." *Philosophical Transactions of the Royal Society* 48: 780–784.

Coulomb, Charles Augustin de. 1779. *Théorie des machines simples*. Paris: De l'Imprimerie Royale.

Coulomb, Charles Augustin de. 1788. "Sur l'électricité et la magnétisme, premier mémoire, construction et usage d'une balance électrique..." *Mémoires de l'Académie Royale des Sciences*. Académie des Sciences, pp. 569-77.

Crookes, William. 1881. "Radiant Matter." *Chemical News* 40: 91-93, 104-107, 127-131.

Descartes, René. 1644. *Principia philosophiae*. Amsterdam: Elzevier.

Einstein, Albert. 1905. "Uber einen die Erzeugung und Verwandlung des Lichtes betreffenden heuristischen Gesichtspunkt." *Annalen der Physik* 17: 132-148.

Faraday, Michael. 1821. "Historical Sketch of Electromagnetism. Part 1." *Annals of Philosophy* 18: 195–200.

Faraday, Michael. 1821–1822. "On Some New Electro-Magnetical Motions, and on the Theory of Magnetism." *Quarterly Journal of Science* 12: 74-96.

Faraday, Michael. 1822. "Historical Sketch of Electromagnetism. Part 2." *Annals of Philosophy* 19: 107–117.

Faraday, Michael. 1832. "On the Induction of Electric Currents and on the Evolution of Electricity from Magnetism." *Philosophical Transactions of the Royal Society* 122: 125–162.

Faraday, Michael. 1839–1855. *Experimental Researches in Electricity and Magnetism*. 3 vols. London: Taylor and Francis (vol. 1, 1839; vol. 3, 1855); London: Richard & John E. Taylor (vol. 2, 1844).

Faraday, Michael. 1852. "On the Physical Character of the Lines of Magnetic Force." In Faraday. *Experimental Researches in Electricity*, vol. 3, 407-437.

Faraday, Michael. 1932-1936. *Faraday's Diary*. 7 vols. London: G. Bell.

Franklin, Benjamin. 1751. *Experiments and Observations on Electricity, made at Philadelphia in America, by Mr. Benjamin Franklin*. London: E. Cave.

Franklin, Benjamin. 1905–1907. *The Writings of Benjamin Franklin*. Edited by Albert H. Smith. 10 vols. New York: Macmillan.

Galvani, Luigi. 1791. *De viribus electricitatis in motu musculari*. Bononiae: Instituti Scientiarum.

Gilbert, William. 1600. *De magnete ... Londini: Excudebat Petrus Short*. [On the Loadstone and Magnetic Bodies]. Translated by Paul Fleury Mottelay. London: B. Quaritch.

Green, George. 1828. *An Essay on the Application of Mathematical Analysis to the Theories of Electricity and Magnetism*. Published at Nottingham in 1828. Reprinted in *Mathematical Papers of the late George Green*, ed. N. M. Ferrers. London: Macmillan, 1871.

Guericke, Otto von. 1672. *Experimenta nova (ut vocantur) magdeburgica de vacuo spatio*. Amsterdam: J. Janssonium a Waesberge.

Hauksbee, Francis. 1705. "Several Experiments on the Attrition of Bodies in Vacuo." *Philosophical Transactions of the Royal Society* 24: 2165–2175.

Hauksbee, Francis. 1709. *Physico-Mechanical Experiments on Various Subjects*. London: R. Brugis.

Heaviside, Oliver. 1892. *Electrical Papers*. 2 vols. London: Macmillan.

Hertz, Heinrich. 1892. *Untersuchungen uber der Ausbreitung der elektrischen Kraft*. Leipzig: A Barth.

Hertz, Heinrich. 1893. *Electric Waves: Being researches on the propagation of electric action with finite velocity Through Space*. Translated by D. E. Jones. New York: Macmillan.

Kant, Immanuel. 1781. *Kritik der Reinen Vernunf*. Riga: Verlegts Johann Friedrich Hartknoch.

Kant, Immanuel. 1787. *Metaphysische Anfangsgründe der Naturwissenschaft*. Riga: Bey Johann Friedrich Hartknoch.

Larmor, J. 1894–1897. "A Dynamical View of the Electric and Luminiferous Medium." *Philosophical Transactions of the Royal Society* (1894) 185: 719–822; (1895) 186: 695–743; (1897) 190: 205–300. Reproduced in J. Larmor, *Mathematical and Physical Papers* (Cambridge: Cambridge University Press, 1929), pt. 1: 1: 414–535; pt. 2: 1: 543–597; pt. 3: 2: 11–132.

Lavoisier, Antoine. 1789. *Traité élémentaire de chimie*. 2 vols. Paris: Cuchet.

Lorentz, H. A. 1892. "La théorie électromagnétique de Maxwell et son application aux corps mouvants." Reprinted in *Collected Papers*. Vol. 2., 164–343. The Hague: Hijhoff, 1934..

Lorentz, H. A. 1895. *Versuch einer Theorie der Electrischen und Optischen Ercheingen*. Leiden: E. J. Brill.

Maxwell, James Clerk. 1856. "On Faraday's Lines of Force." Reprinted in *The Scientific Papers of James Clerk Maxwell*, vol. 1, 155-229.

Maxwell, James Clerk. 1861-62. "On Physical Lines of Force." Reprinted in *The Scientific Papers of James Clerk Maxwell*, vol. 1, 451-513.

Maxwell, James Clerk. 1865. "A Dynamical Theory of the Electromagnetic Field." *Philosophical Transactions of the Royal Society of London* 155: 459-512. Reprinted in *The Scientific Papers of James Clerk Maxwell*, vol. 1, 526-597.

Maxwell, James Clerk. 1873. *A Treatise on Electricity and Magnetism*. 2 vols. Oxford: Clarendon.

Maxwell, James Clerk. 1890. *The Scientific Papers of James Clerk Maxwell*. 2 vols. Edited by W. D. Niven. Cambridge: Cambridge University Press.

Michelson, Albert. 1881. "The Relative Motion of the Earth and the Luminiferous Ether." *American Journal of Science* 22: 120–129.

Michelson, Albert, and Edward Morley. 1887. "On the Relative Motion of the Earth and the Luminiferous Ether." *American Journal of Science* 34: 333–345.

Michell, John. 1750. *A Treatise of Artificial Magnets*. Cambridge: J. Bentham.

Musschenbroek, Pieter van. 1729. *Dissertatio physica experimentalis de magnete*. Vienna: Trattner.

Newton, Isaac. 1687. *Philosophia Naturalis Principia Mathematica*. London: Jussu Societatis Regiæ ac Typis Josephi Streater; prostat apud plures Bibliopolas. Translated by Andrew Motte. Revised by Florian Cajori. *Sir Isaac Newton's Mathematical Principles of Natural Philosophy and His System of the World*. Berkeley: University of California Press, 1934.

Newton, Isaac. 1704. *Opticks, or, A treatise of the reflexions, refractions, inflexions and colours of light*. London: S. Smith. and B. Walford.

Newton, Isaac. 1959–77. *The Correspondence of Isaac Newton*. Eds. H. W. Turnbull, J. F. Scott, A. and Rupert Hall. 7 volumes. Cambridge: The University Press, 1959–1977.

Nicholson, William, ed. 1800. *Journal of Natural Philosophy, Chemistry, and the Arts*. London: Printed by W. Stratford, Crown-Court, Temple-Bar for the Author.

Nollet, Jean-Antoine. 1746. *Essai sur l'électricité des corps*. Paris : Chez les Freres Guerin

Nollet, Jean-Antoine. 1749. *Recherches sur les causes particulieres des phénoménes électriques*. Paris : Chez les Freres Guerin.

Norman, Robert. 1596. *The Newe Attractive, Containing a Short Discourse of the Magnes or Loadstone* ... London: n.p.

Oersted, Hans Christian. 1813. *Recherches sur l'identité des forces chimiques et électriques*. Traduit de l'allemand par Marcel de Serres. Paris: J. G. Dentu.

Oersted, Hans Christian. 1820. "Experimenta circa effectum conflictus electrici in acum magneticum." Reprinted in *Annals of Philosophy* 16 (1820), 273-276.

Ohm, Georg Simon. 1827. *Die Galvanische kette*. Berlin: T. H. Riemann.

Peregrinus, Petrus. 1269/1902. *Petrus Peregrinus de Maricourt and his Epistola de Magnets*. Translated by Silvanus P. Thompson. London: Charles Whittingham.

Priestley, J. 1767. *History and Present State of Electricity, with Original Experiments*. London: printed for J. Dodsley, J. Johnson, B. Davenport, T. Cadell.

Röntgen, Wilhelm. 1896. "A New Kind of Rays." *Nature* 53: 272–276.

Rutherford, Ernest. 1899. "Uranium Radiation and the Electrical Conduction Produced by It." *Philosophical Magazine* 47: 109–163.

Rutherford, Ernest. 1904. *Radio-Activity*. Cambridge: Cambridge University Press.
Thomson, J. J. 1897. "Cathode Rays." *Philosophical Magazine* 44: 293–298
Thomson, J. J. 1936. *Recollections and Reflections*. London: G. Bell and Sons.
Thomson, William. 1882–1911. *Mathematical and Physical Papers*. Cambridge: Cambridge University Press. 6 vols.
Volta, Alessandro. 1800. "On the Electricity Excited by the Mere Contact of Conducting Substances of Different Kinds." *Philosophical Transactions of the Royal Society* 90: 403–431.

SECONDARY SOURCES

Buchwald, Jed Z. 1985. *From Maxwell to Microphysics: Aspects of Electromagnetic Theory in the Last Quarter of the Nineteenth Century*. Chicago: University of Chicago Press.
Buchwald, Jed Z. 1994. *The Creation of Scientific Effects: Heinrich Hertz and Electric Waves*. Chicago: University of Chicago Press.
Campbell, L., and W. Garnett. 1882. *The Life of James Clerk Maxwell*. New York: Macmillan.
Caneva, K. L. 1997. "Physics and *Naturphilosophie:* A Reconnaissance," *History of Science* 35: 35–107.
Clark, David H. 2001. *Newton's Tyranny : The Suppressed Scientific Discoveries of Stephen Gray and John Flamsteed*. New York: W. H. Freeman.
Cohen, I. Bernard. 1956. *Franklin and Newton*. Philadelphia: American Philosophical Society.
Cohen, I. Bernard. 1990. *Benjamin Franklin's Science*. Cambridge, Mass.: Harvard University Press.
Darrigol, Olivier. 2000. *Electrodynamics from Ampère to Einstein*. Oxford: Oxford University Press.
Ennemoser, Joseph. 1854. *The History of Magic*. 2 vols. Bohn: H. G. Bohn.
Fahie, J. J. 1918. "Galileo and Magnetism: A Study on Lodestones." *Journal I.E.E.* (British) 273: 246–49.
Gower, B. 1973. "Speculation in Physics: The History and Practice of Naturphilosophie," *Studies in History and Philosophy of Science* 3: 301-356.
Grant, Edward. 1981. "Peter Peregrinus." *Dictionary of Scientific Biography*. Ed. Charles Coulston Gillispie. New York: Charles Scribner's Sons, vol 10, 532–540.
Harman, Peter 1971. "Faraday's Theories of Matter and Electricity." *The British Journal for the History of Science* 5: 235–257.
Harman, Peter. 1998. *The Natural Philosophy of James Clerk Maxwell*. Cambridge: Cambridge University Press.

Heilbron, J. L. 1979. *Electricity in the 17th and 18th Centuries: A Study of Early Modern Physics.* Berkeley: University of California Press.

Heilbron, J. L. 1981. *The Effluvial Theory of Electricity.* New York: Arno Press.

Heimann, P. M. 1969–1970. "Maxwell and the Modes of Consistent Representation." *Archive for History of Exact Science* 6: 171–213.

Hesse, Mary. 1961. *Forces and Fields: The Concept of Action at a Distance in the History of Physics.* Edinburgh: T. Nelsen.

Hofmann, James R. 1987. "Ampère, Electrodynamics, and Experimental Evidence." *Osiris* 3: 45–76.

Home, R. W. 1992. *Electricity and Experimental Physics in Eighteenth Century Europe.* Brookfield, Vt.: Ashgate.

Hunt, B. J. 1991. *The Maxwellians.* London: Cornell University Press.

McCormmach, Russell. "H. A. Lorentz and the Electromagnetic View of Nature." *Isis* 61: 459–497.

Meyer, Herbert W. 1971. *A History of Electricity and Magnetism*, Cambridge, Mass.: MIT Press.

Mottelay, Paul F. 1922. *Bibliographical History of Electricity and Magnetism.* London: C. Griffin.

Pancaldi, Giulano. 2003. *Volta: Science and Culture in the Age of the Enlightenment.* Princeton, N.J.: Princeton University Press.

Park, Benjamin. 1895. *The Intellectual Rise in Electricity.* London: Longmans, Green & Co.

Roller, Duane H. 1959. *The De Magnete of William Gilbert.* Amsterdam: M. Hertzberger.

Tait, Peter. 1873. "Clerk-Maxwell's Electricity and Magnetism." *Nature* 7: 478–480.

Warwick, Andrew. 1993. "Frequency, Theorem and Formula: Remembering Joseph Larmor in Electromagnetic Theory." *Notes and Records of the Royal Society of London* 47: 49–60.

Whittaker, E. T. 1951. *A History of the Theories of Aether and Electricity: From the Age of Descartes to the Close of the Nineteenth Century.* 2 vols. New York: Nelson.

Williams, L. Pearce. 1966. *The Origins of Field Theory.* New York: Random House.

Woodruff, A. E. 1962. "Action at a Distance in Nineteenth Century Electrodynamics." *Isis* 53: 439–459.

INDEX

Académie des Sciences, 30–31, 43, 45, 68–69, 72
Action at a distance, 2–3, 23, 86–87, 94–95, 99–100, 108
Advancement of Learning (Francis Bacon), 14
Aepinus, Franz (1744–1802), 40
Air pump, 20, 117, 125
Alchemy, 14
Alexander of Aphrodisias (fl. A.D. 200), 3
Ampère, André Marie (1775–1836), 68–71, 76, 79, 81, 88
Ampère's Law, 72, 101, 103
Anion, 86
Anode, 86
Arago, Dominique François (1786–1853), 68–69, 71, 79, 81
Arago's disk (Arago's wheel), 81–82
Aristotle (384 B.C.–322 B.C.), 10
Armstrong, Edwin H. (1890–1954), 115
Astatic needle, 69
Atom, 54, 59–60, 119; atomic number, 132; atomic weight, 86; disintegration of, 133; models of, 126, 133–36; theory of, 84
Atomism: hypothesis, 132; Kant's critique of, 63

Bacon, Francis (1561–1626), 13
Bain, Alexander (1810–1877), 113
Barentz, Wilhelm (d. 1597), 18 n.2
Barlowe, William (d. 1568), 9
Becquerel, Antoine Henri (1852–1908), 129–30, 133
Beddoes Pneumatic Institution, 56
Beeckman, Isaac (1588–1637), 15
Bell, Alexander Graham (1847–1922), 113
Bennet, Abraham (1750–1799), 33
Berlin Academy of Science, 105
Berzelius, Jöns Jacob (1779–1848), 58–59
Biot, Jean-Baptiste (1774–1862), 60, 67
Biot-Savart Law, 72
Bohr, Niels Hendrik David (1885–1962), 135, model of the atom, 136
Bonaparte, Napoleon (1769–1821), 55
Boyle, Robert (1627–1691), 19–21, 24
Broglie, Louis de (1892–1987), 136

Cambridge Philosophical Society, 96
Canton, John (1712–1782), 33
Cardano, Girolamo (1501–1576), 11–12
Carlisle, Anthony (1768–1840), 56

Cathode (negative electrode), 86, 118–19, 121; radiation, 118
Cation, 86
Cavallo, Tiberius (1749–1809), 33
Cavendish, Henry (1731–1810), 45, 109
Cavendish Laboratory, 94, 124, 131
Certain Physiological Essays (Robert Boyle), 19
Chemicals: affinity of, 56, 58–59, 65, 87; decomposition of, 85
Collinson, Peter (c. 1693–1768), 37–38
Comte, Auguste (1798–1857), 64
Condenser (capacitor), 29, 30–31, 45, 103
Conduction of Electricity Through Gases (J. J.Thomson), 124
Conductor, 39, 42, 45, 70, 81, 83–85, 87, 109
Conservation of electric charge, 39, 41
Conservation of energy, law of, 101
Coulomb, Charles Augustin (1736–1806), 7, 33, 40, 43–45; 59–60, 66–67, 70, 88,
Coulomb's Law, 44, 72, 96
Critik der Reinen vernunft (Immanuel Kant), 63
Crookes, William (1832–1919), 118–20, 132
Cruickshank, William, 56
Curie, Marie Sklodowska (1867–1934), 130–31
Curie, Pierre (1859–1906), 130–31
Cuvier, Georges (1769–1832), 79

Dally, Clarence M., 123
Darwin, Charles, 7
Davy, John (1790–1868), 84
Davy, Humphry (1778–1829), 56, 60–61, 75–80
De Anima (Aristotle), 1–2,
De Magnete (William Gilbert), 6, 8–15
De Revolutionibus (Nicolas Copernicus), 13
Desaguliers, John (1683–1744), 27
Descartes, René (1596–1650), 8, 14–19, 29–30

"De viribus electricitatis in motu musculari commentarius" (Luigi Galvani), 50–51
Die galvanische Kette, mathematisch bearbeitet (Georg Simon Ohm), 73
Dielectric, 103, 106
Digges, Thomas (1546–1595), 11, 13
Dirac, Paul (1902–1984), 136
Disintegration hypothesis, 133
Displacement current, 103–4
Doctrine of the unity of forces of nature, 65, 68
Dufay, Charles François de Cisternay (1698–1739), 11, 36, 70
"Dynamical Theory of the Electromagnetic Field" (James Clerk Maxwell), 99, 109
"Dynamical Theory of the Electric and Luminiferous Medium" (Joseph Larmor), 110
Dynamo, 84, 90; disk, 83; fluid, 83; theory of, 83

Edison, Thomas Alva (1847–1931), 90–91, 122
Edison and Swan United Electric Light Company, 90
Einstein, Albert (1879–1955), 101, 135; theory of the photoelectric effect, 135
Electrician, 73
Electricity: alternating current (AC), 90; animal, 49–50, 52, 74 n.1, 84; battery, electric, 51, 53–56, 76; charge, electric, 11, 30–31, 33, 38–41, 46–47, 49, 51–52, 59, 61, 70, 77, 86, 95–96, 101, 103, 118, 124–25, 135; circuit, electric, 57, 69, 72, 81–82, 106; conduction, electrical, 22, 25–29, 35, 124; conductor, electrical, 27, 57, 83; current, electric, 45, 49, 52–59, 61, 66–74, 76–82; 86, 90–91, 96–99, 110, 117, 121, 124–25, 134; de-composition, electrical, 55–56; direct current (DC), 91; discharge, electric, 117; effluvia, electrical, 3,

11, 24, 35, 37; energy, electrical, 83; field, electric, 101, 103–5, 118, 121, 125, 149; fluid, electric, 11, 34–40, 46–47, 51, 54–55, 57, 68, 86–88; force, electric, 100, 108, 135; induction, electrical, 22, 106; insulator, electrical, 27, 30, 36; intensity, electrical, 96; luminescence, 25; machine, electrical, 22, 21–32, 38, 50, 83; magnetoelectricity, 85; motor, electric, 76, 91; particle, electric, 99; polarity, electrical, 87; polarization, electric, 99–100, 110; potential, electric, 45, 54; power, electrical, 91; powers, electrical, 88; power stations, electrical, 90; quantity of, 72, 86; quantum of, 86; resinous, 36, 39; static, 1, 42, 49, 51–53, 60, 71, 81, 85; tension, electrical, 54, 72, 84, 87, 97; thermoelectricity, 85; vitreous, 36, 39; voltaic, 85; wave, electric, 81, 101, 103, 107–8
Electric Waves (Heinrich Hertz), 109
Electrochemical decomposition, 85–86, 88
Electrochemistry, 55–60, 86
Electrode, 86
Electrodynamic molecule, 71
Electrodynamics, 49, 70–71, 110
Electrolysis, 58, 86
Electrolyte, 57, 86, 125
Electromagnet, 68, 72, 81, 88
Electromagnetism, 67–69, 75, 81, 136: effects, 80; fields, 95, 99, 101, 103, 109, 111–12; forces, 77, 103; induction, 77, 81–83; lines of force, 94–98; potential, 94; processes, 99; radiation, 100, 107–8, 135; rotations, 75–77, 79–80; theory, 76, 100–104, 108, 110–11; waves, 99–100, 103–9
Electrometer, 33, 52–53
Electromotive force (EMF), 57, 72
Electron, 1, 54, 61, 86–87, 112, 119, 126–27, 131–36; measurement of charge of, 134; theory of, 110–13, 126; tube, 114
Electrophorous, 33, 51

Electroscope, 33, 51
Electrostatics: charge, 86; field, 125; generator, 78, 85; induction, 33, 87; machine, 84
Electrotonic state, 84, 87, 96
Empedocles of Acragas (c. 490 B.C.–c. 430 B.C.), 2–3; 11
"Epistola de Magnete" (Petrus Peregrinus) 4, 33
Energy, 89–90; mechanical, 90
"Essay on the Application of Mathematical Analysis to the Theories of Electricity and Magnetism" (George Green), 95
Ether, 15, 23–24, 89, 95, 99–100, 102, 109–13, 118, 133
Ether drag hypothesis, 111–12
Ether wind, 111
"Experimenta circa effectum conflictus electrici in acum magneticum" (Hans Christian Oersted), 19
Experimental Researches in Electricity (Michael Faraday), 94, 97, 117
Experimenta Nova (Otto von Guericke), 22
Experiments and Notes About the Mechanical Origine or Production of Electricity (Robert Boyle), 20

Faraday, Michael (1791–1867), 45, 61, 75–89, 95–97, 100, 104, 117, 124
Faraday cylinder, 124
Faraday rotation, 89, 97
Faraday's dark space, 117
Faraday's law of induction, 101
Faraday's laws of electrolysis, 85, 104
Fessenden, Reginald Aubrey (1866–1932), 114–15
Field theory, 99–100, 111
Fish-eye lens, 93
FitzGerald, G. F. (1851–1901), 102, 112, 126
FitzGerald-Lorentz contraction, 112–13
Fizeau, Armand Hippolyte (1819–1896), 103, 110
Flamsteed, John (1646–1719), 26
Fluorescence, 117, 121

Fluoroscope, 123
Foucault, Jean Léon (1819–1868), 103
Fourier, Joseph, 72
Fracastoro, Girolamo (1478–1533), 11
Franklin, Benjamin (1706–1790), 11, 33, 36–42, 53, 60, 70
Fresnel, Augustin Jean (1788–1827), 71, 102, 111
Fresnel drag coefficient (ether drag hypothesis), 112

Galilei, Galileo (1564–1642), 8, 14
Galvani, Luigi (1737–1798), 49–51
Galvanism, 49, 51, 57, 60, 65; galvanic trough, 56; galvanic battery, 76, 78
Galvanometer: 82, 84–85, 106; astatic, 74 n.1; principle of, 66
Gas discharges, 121
Gauss's law, 101
Gauss's theorem, 95
Gay-Lussac, Joseph-Louis (1778–1850), 79
Geissler, Heinrich (1814–1879), 117
Gilbert, William (1540–1603), 8–18, 21–22
Goldstein, Eugen (1850–1930), 118–20
Gravitation, 88–89, 108, 136; central forces, 77, 80; field, 109; theory of, 16, 47; universal, 23, 63
Gray, Stephen (1670–1736), 25–28, 33–36
Green, George (1793–1841), 94
Green's theorem, 95
Guericke, Otto von (1602–1686), 20–21

Halley, Edmond (1656–1742), 26
Hauksbee, Francis (1660–1713), 24–25, 27, 28 n.1, 117
Heat: flow, 95–96; theory of, 95
Heaviside, Oliver (1850–1925), 102, 110
Heisenberg, Werner Karl (1901–1976), 136; uncertainty principle, 135
Helmholtz, Hermann von (1821–1894), 86, 101, 105–6
Henry, Joseph (1797–1878), 83, 90

Hertz, Heinrich Rudolf (1857–1894), 101, 103, 105–9, 113, 118, 125
Hippocrates (c. 460 B.C.–c. 370 B.C.), 10
Historia Coelestis Britannica (John Flamsteed), 26
Hooke, Robert (1635–1703), 20
Hopkinson, Thomas (1709–1751), 38
Humboldt, Alexander von (1769–1859), 79

Induction coil, 106
Inertia, law of, 12
Interferometer, optical, 111
Institut de France, 68
Insulators, 40, 41, 87
Inverse square law, 42, 45
Invisible college, 20
Isotopes, 132–33

Kant, Immanuel (1724–1804), 63, 68
Kinnersley, Ebenezer (1711–1778), 38
Kircher, Athanasius (1602–1680), 18 n.1
Kirchhoff, Gustav Robert (1824–1887)
Kleist, Ewald von (1715–1759), 29

Lagrange, Louis (1736–1813), 47
Laplace, Pierre Simon (1749–1827), 47, 60, 95
Larmor, Joseph (1857–1942), 102, 110, 126
Lavoisier, Antoine Laurent (1743–1794), 7, 56, 58, 78
Lenard, Philip (1862–1947), 118, 126
Leyden jar, 30–31, 36–38, 40–41, 50–53, 82
Light: dispersion of, 102; electric, 90; as an electromagnetic disturbance, 100, 103–4; Fresnel's theory of, 102; as longitudinal vibrations of the ether, 102, 111, 123; polarized, 4, 88, 97, 99, 102; reflection of, 101–2; refraction of, 99, 101–2; as transverse vibrations of the ether, 98–99, 102, 111, 123; velocity of, 98, 104,

111; waves of, 105, 108, 109, 135; wave theory of, 60, 71, 99, 102, 111
Lightning, 40–42, 50, 66, 115
Lodge, Oliver (1851–1940), 102, 108, 133
Lorentz, Hendrik (1853–1928), 102, 110–12, 126

Magnes Siue de Arte Magnetica (Athanasius Kircher), 18
Magnet: bar, 119; horseshoe, 82, 119; magnetite, 2; permanent magnet, 71, 80, 82; theory of, 40, 70
Magnetism: charge, 101; compass, 3–5, 16, 43, 67–69, 75, 77, 80–81, 85; diamagnetic, 89; dip, 9, 12; excitation, 89; field, 69, 71–72, 81–83, 88–89, 91, 97, 101, 104, 119, 125; fluid, 46, 68; force, 89; induction, 15, 96; influence, 3; lines of force, 89, 101; paramagnetic, 89; philosophy, 14; polarity, 74 n.1; polarization, 99–100; pole, 4, 12, 75–77, 89; variation, 12
Marconi, Marchese Guglielmo (1837–1937), 108, 113
Marconi's Wireless Telegraph Co. Ltd., 114
Maxwell, James Clerk (1831–1879), 45, 93–101, 104–6, 108–9, 111, 123
Maxwell's equations, 93, 99–100, 102–3, 108–9, 111
Mechanical philosophy, 19
Mechanization of nature, 19
"Mémoire sur l'action mutuelle de deux courants électriques" (André Marie Ampère), 69
"Mémoire sur la thérie mathématique des phénomènes électrodynamique, uniquement déduite de l'expérience" (André Marie Ampère), 71
Metaphysiche Anfangsgründe der Naturwissenschaft (Immanuel Kant), 64
Mitchell, John (1724–1793), 42
Michelson, Albert A. (1852–1931), 111

Michelson-Morley experiment, 112
Millikan, Robert Andrews (1868–1953), 134
Model, 87; geometrical, 96; mechanical, 97, 110; of the ether, 102, 110; phenomenological, 83; plum pudding (raisin cake), 127; provisional, 84
Moll, Gerritt (1785–1838), 81
Morley, Edward (1838–1923), 111
Morse, Samuel F. B. (1791–1872), 113
Morse code, 113–14
Musschenbroek, Pieter van (1692–1761), 29, 32

Naturphilosophie, 65
Navigiator's Supply (William Barlowe), 9
Neumann, Carl (1832–1925), 106
Neutron, 127
Newe Attractive (Robert Norman), 8
New Experiments Physico-Mechanical, Touching on the Spring of Air and Its Effects (Robert Boyle), 20
Newton, Isaac (1642–1727), 7–8, 23–29, 37, 41, 43, 63, 67, 70, 100; rules of reasoning, 63
Nicholson, William (1809–1845), 56
Nobili, Leopold (1784–1835), 74 n.1
Nollet, Jean-Antoine Abbe (1700–1770), 31, 40
Norman, Robert (late sixteenth century), 8
"Notes on the Electromagnetic Theory of Light" (James Clerk Maxwell), 102

Oersted, Hans Christian (1777–1851), 65–67, 74–76, 82
Ohm, Georg Simon (1787–1854), 45, 72–73
Ohm's Law, 73
"On Faraday's Lines of Force" (James Clerk Maxwell), 95
"On Physical Lines of Force" (James Clerk Maxwell), 97
"On some new Electro-Magnetical Motions, and on the Theory of Magnetism" (Michael Faraday), 76

"On the Electricity Excited by the Mere Contact of Conducting Substances of Different Kinds" (Alessando Volta), 55
"On the Physical Character of the Lines of Force" (Michael Faraday), 89
"On the Physical Character of the Lines of Magnetic Force" (Michael Faraday), 97
Opticks (Isaac Newton), 23
Opus Tertium (Roger Bacon), 6
Oscillator, 107, 114–15
Outlines of Experiments and Enquiries Respecting Sound and Light (Thomas Young), 60

Pearl Street Station, 90
Pemberton, Henry (1694–1771), 37
Pepys, William Hasledine (1775–1856), 57
Peregrinus, Petrus (c. 1200–c. 1270), 3–6; 8; 10
Perfect Description of the Caelestiall Orbes (Thomas Digges), 13
Perrin, Jean Baptiste (1870–1942), 124
Phosphorescence, 129
Photoelectricity, 135
Photon, 136
Pitchblende, 130
Planck, Max (1858–1947), 135
Planck's constant, 135
Plücker, Julius (1801–1868), 117
Poisson, Siméon Denis (1781–1840), 47, 60, 72, 81, 88
Poor Richard's Almanac (Benjamin Franklin), 37
Positron, 136
Priestley, Joseph (1733–1804), 41
Proton, 1, 54, 127, 132–34, 136
Philosophia Naturalis Principia Mathematica (Isaac Newton), 22–24, 26, 37, 63

Quantum electrodynamics, 136
Quantum energy, 135
Quantum mechanics, 136
Quantum theory, 101, 135

Radioactivity, 129–33; and half-life, 132; measurement of, 132
Radio-Activity (Ernest Rutherford), 133
Radiography, 122
Radios: antennas, 106; portable, 115
Radio transmission, 114–15
Radio waves, 105, 108
Raleigh, Lord, 123
Rays: alpha, 125, 131; beta, 131; canal, 118; cathode, 113, 117–20, 124, 126; gamma, 131, 133; penetrating, 122; radium, 134; ultraviolet, 122; of uranium, 130; X-rays, 119–24, 130–31, 133–34
Receiver technology: regenerative, 114; superheterodyne, 114
Recent Researches in Electricity and Magnetism (J. J. Thomson), 124
"Recherches sur l'identité des forces chimiques et électriques" (Hans Christian Oersted), 65
Resistance, 73
Röntgen, Wilhelm Konrad (1845–1923), 120–24, 129
Royal Danish Society of Science, 66
Royal Institution (Great Britain), 57, 60–61, 78–80, 126
Royal Society of London, 20, 24–28, 32, 45, 68, 79–80, 124
Ruhmkorff coil, 120
Rumford, Count (Benjamin Thompson) (1753–1814), 57
Rutherford, Ernest (1871–1937), 124–26, 131–33, 135; model of the atom, 133–36

Salvioni, E. (1858–1920), 123
Saturn's rings, 93
Saussure, Horace Benedict de (1740–1799), 33
Savart, Felix (11791–1841), 72
Schrödinger, Erwin (1887–1961), 136

Seebeck, Thomas Johann, 73
Shelling, F.W.J. (1775–1854), 65
Sloan, Hans (1660–1753), 26–28
Soddy, Frederick (1877–1956), 132
Solenoid, 81
Static, 115
Stokes, G. G. (1819–1903), 111
Stoney, George Johnstone (1826–1911), 126
Sülzer, Johann Georg (1720–1779), 52
Swan, Joseph Wilson (1828–1914), 90
Syng, Philip (1703–1789), 38

Tait, Peter (1831–1901), 94
Tentamen Theoriae Electricitatis et Magnetismi (Franz Aepinus), 40, 47
Tesla, Nikola (1856–1943), 93
Thales (c. 640–546 B.C.), 1
Theorie des Machines Simples (Charles Auguste de Coulomb), 43
Theory of special relativity, 113
Thermocouple, 73, 84
Thomson, Joseph John (1856–1940), 61, 109, 123–26
Thomson, William (Lord Kelvin) (1824–1907), 88, 95, 96–97; molecular vortex theory, 97–99; transmutation of the elements, 133
Torsion balance, 43–44, 46
Traité Elementaire de Chimie (Antoine Lavoisier), 56
Transformer, 81
Transistor, 115
Transmutation of the elements, 133
Treatise on Electricity (James Clerk Maxwell), 94, 102, 109–10

Tube: cathode ray, 118; Crookes, 119–20, 129

"Über eine neue Art von Strahlen" (Wilhelm Rontgen), 120
"Uranium Radiation and the Electrical Conduction Produced by It" (Ernest Rutherford), 131

Vacuum, 20–21, 87, 117, 119
Varley, Cromwell (1828–1883), 118–19, 124
Versorium, 10, 12, 32
Villard, Paul-Ulrich (1860–1934), 131
Von Kleist, Ewald G. (1700–1748), 29
Volta, Alessandro (1745–1827), 33, 40, 46, 51. 84
Voltage, 53–54, 73, 82–83, 106, 117
Volta's pile, 53–60, 73, 78, 82, 84, 86

Watson, William (1715–1787), 32
Wave: equations and, 104, frequency, 107; Hertzian, 108; sound, 114; velocity of, 107
Weber, Wilhelm Eduard (1804–1891), 104, 106
Westinghouse, George (1846–1914), 91
Whewell, William (1794–1866), 86
Wireless Telegraph and Signal Co., Ltd., 114
Wireless telegraphy, 113, 135
Wollaston, William Hyde (1766–1828), 77, 79–80, 85

Young, Thomas (1773–1829), 60

About the Author

BRIAN BAIGRIE is Associate Professor in the Institute for History and Philosophy of Science at the University of Toronto. He is the author of many articles in history and philosophy of science, and has edited *Scientific Revolutions: Primary Texts in the History of Science, History of Modern Science and Mathematics* and *Life Scientists of the Twentieth Century*.

QC
507
.B35

2007